SYSTEMS
DEVELOPMENT
A Practical Approach

SYSTEMS DEVELOPMENT
A Practical Approach

William Amadio
Rider College

Mitchell Publishing, Inc.
Innovators in Computer Education
915 River Street • Santa Cruz, California 95060
(800) 435-2665 • In California (408) 425-3851
"A Random House Company"

Desktop Publishing Service: C. Morales, Arizona Publication Service
Cover Design: Nancy Benedict
Printer: R. R. Donnelley & Sons
Product Development: Raleigh S. Wilson
Production Management: Greg Hubit Bookworks
Sponsoring Editor: Erika Berg
Text Design: Nancy Benedict

Printed in the United States of America
10   9   8   7   6   5   4   3   2   1

Library of Congress Card Catalog No.: 88-61872

ISBN: 0-394-39232-9 (text only)
      0-394-39430-5 (text and software)

For Bill, Jr.—
you're the best!

CONTENTS

PART FOUR
Completing the Systems
Development Life Cycle   278

▶ **CHAPTER 12**
Systems Implementation   280

▶ **CHAPTER 13**
Completing the
Implementation   304

▶ **CHAPTER 14**
A Critical Appraisal of the
Horatio Systems   320

PREFACE

To the Student

This text deals with the analysis, design, implementation, and maintenance of computer-based information systems in business organizations. These topics are of concern to *all* business professionals regardless of their field of specialization.

In reading this text, you will discover that the systems development process is made up of technical, economic, social, and political components. No one component may be considered to the exclusion of the others. If you are an information systems major, pay particular attention to the non-technical issues. If your major is outside of information systems, try to appreciate the importance of the technical issues to the process. When all parties recognize and understand competing points of view, a successful project results.

I am happy you decided to take this course; I hope you find it interesting and worthwhile. Good luck!

To the Instructor

Every instructor who regularly teaches systems analysis and design knows the meaning of the phrase "love-hate relationship." We know the importance, the beauty, and the excitement of the subject, but we also know the difficulty of teaching it and the frustration that comes when our results do not match our expectations.

I believe the source of our difficulty is articulated in Kolb's Experiential Learning Model, which states that the learning/problem solving process is "both active and passive, concrete and abstract. It [the process] can be conceived of as a four-stage cycle: (1) concrete experience is followed by (2) observation and reflection which leads to (3) the formation of abstract concepts and generalizations which lead to (4) hypotheses to be tested in future action which in turn leads to new experiences."[1]

---

[1] D. Kolb, I. Rubin, and J. McIntyre. *Organizational Psychology* (Englewood Cliffs, NJ: Prentice-Hall, 1974), p. 28.

The difficulty of teaching the systems course lies in the fact that most undergraduate students do not bring concrete systems experience to the course; they usually bring programming experience. The stage hypothesis of Kolb's model would suggest, then, that the systems instructor's first task is to expand the students' programming experience into concrete systems experience. Otherwise, the instructor is beginning the learning cycle at stage 2, which translates into trying to teach students to analyze, design, implement, and maintain something they have never seen.

I believe that the combination of microcomputer technology, fourth-generation languages (4GLs), and computer-aided software engineering (CASE) tools provides the means to implement the full learning cycle in the systems course. This technology can be exploited to create a new learning environment in which students acquire task-specific and critical thinking skills that are transferable to any hardware/software work environment.

Two complete, microcomputer-based systems accompany this text. One of them provides a meaningful first experience with an information system operating within the context of a business organization. With this common base of experience, the classroom becomes a place for students and instructor to observe, reflect, conceptualize, and generalize together. The learning cycle is complete when the second system is used to provide an arena for active experimentation with the newly developed concepts and skills.

This *do it—study it—do it again* approach is fundamental to the design of this text.

Audience

Systems Development: A Practical Approach is designed for a first course in systems development. This course is usually offered at the sophomore and junior levels by both two-year and four-year institutions.

Organization

The text is organized into five parts. Part One provides an overview of systems development practice and introduces the Horatio & Co. Cost Control System, a working system provided on diskette with this text. The software development environment is dBASE III PLUS supplemented with GENIFER, a CASE tool that provides a data dictionary, report writer, screen painter, and dBASE code generator.

GENIFER proved to be just the right tool to bring a working system into the classroom without getting bogged down in programming details.

GENIFER, dBASE, and cost control system screens are incorporated directly into the text to support a *generic* discussion of system concepts. **Student and instructor familiarity with dBASE III PLUS and GENIFER is not assumed or necessary.**

Parts Two through Four of the text look back in time to reconstruct the development of the cost control system from problem analysis through integration of the system with the existing infrastructure of the organization.

Part Five examines technical and organizational issues not covered in the development of the cost control system and provides a look at trends that are likely to affect systems development practice in the future.

After an overview of systems development practice is presented in Chapter 1, the Kolb four-stage model is incorporated into the structure of the remaining chapters of Parts One through Four. Each chapter begins with a concrete experience with the Horatio & Co. Cost Control System. These experiences include using the software, reconstructing analysis and design sessions, preparing deliverables, and reviewing decisions.

The latter part of each chapter reflects on the new experience and generalizes to the appropriate concepts. Each chapter concludes with a set of thought questions designed to stimulate further observation, reflection, and conceptualization and a set of exercises that asks the student to apply the newly learned concepts to the development of the Horatio & Co. Credit Union System, a second working system supplied on diskette with this text. The *Instructor's Manual* accompanying the text contains complete answers to all 78 thought questions and 94 exercises.

In making choices about conceptual content, I relied on the model curricula of DPMA and ACM and the work of Vitalari.[2] Any shortcomings, of course, rest with me.

More often than not, the difficult choices involved not *what* to cover but rather *when* to cover it. Because of the close connection to a real system, topics like feasibility, organizational structure, management approval, and effective communication made their presence felt in each and every chapter.

To help with your assessment of the content, I have prepared the accompanying analysis, which lists the dominant theme(s) of each chapter, and representative figures that communicate the flavor of some of the treatments.

[2]See N. Vitalari, "Knowledge as a Basis for Expertise in Systems Analysis: An Empirical Study," *MIS Quarterly* 9, (September, 1985): 221–40.

Dominant Chapter Themes	Representative Figures
Part One: The Foundation of This Text	
1. Overview	1.2, 1.3
2. System outputs: reports and inquiries	2.6, 2.13
3. System inputs: maintenance processes	3.4, 3.5, 3.16, 3.17
Part Two: Systems Analysis	
4. Objectives Tactics Information system functions	4.2, 4.4
5. DFD-based modeling Requirements Communication techniques	5.11, 5.23, 5.25, 5.26
6. Assessing benefits Boundaries: system wide automated/human work mix	6.1, 6.6, 6.11
7. Feasibility: operational technical economic Buying software Building software: prototyping detailed design	7.1, 7.4, 7.5, 7.6
Part Three: Systems Design	
8. Management approval Project management Design: hardware procedures personnel	8.2, 8.3, 8.4
9. Design: data software	9.13, 9.26, 9.28, 9.30
10. Inside a CASE-developed system	10.4, 10.7, 10.8, 10.14
11. Evaluation: refining prototypes team reviews	11.1, 11.8
Part Four: Completing the Systems Development Life Cycle	
12. Implementation and integration	12.1, 12.5, 12.6, 12.8
13. Ongoing documentation	13.1
14. A critical review of the Horatio systems	14.1, 14.2

Highlights

Several features of this text deserve special mention.

▶ **New learning environment.** This text presents an organized set of learning experiences built around two microcomputer-based application systems that are supplied on diskette. The text first introduces the students to the systems as users. Subsequently, they learn the principles of analysis and design through a reconstruction of the development of these systems. Every phase of the life cycle is relevant because the students have seen and used the finished product and can evaluate development activities in terms of their own experience.

▶ **Active use of CASE technology.** The Horatio & Co. Cost Control System was developed with dBASE III PLUS supplemented by the GENIFER CASE tool. In Chapter 10, students use GENIFER's documentation database to learn the inner workings of the cost control system in the same way they would use any CASE design database in practice. The end-of-chapter exercises ask students to construct CASE specifications for the credit union system. Thus, there is no need to "cover" CASE technology. Students use it actively in their pursuit of the text material.

▶ **Expanded systems development life cycle.** Traditional evaluation of alternatives is expanded to include choice of software development methodology. Prewritten software packages, prototyping, and detailed design are presented as a portfolio of methods to be used according to the circumstances of the project. The use of each method is integrated into the unifying framework of the expanded life cycle.

▶ **Full spectrum of information technology.** The Horatio & Co. systems cover a range of applications from transaction processing through decision support. From the very outset, these systems are discussed within a broad context of information technology that includes expert systems, office automation, telecommunications, and factory automation. Part

Five discusses the dominant issues faced in the analysis, design, and implementation of these newer technologies.

▶ **Higher-level conceptual treatment.** The specific-to-general approach that dominates the text is exploited several times to reach conceptual levels not normally associated with this course. For example, using the Horatio & Co. organization as a base, Part Five reviews the McKenney & McFarlan model of technology assimilation and discusses its implications for systems analysis and design. The concrete experiences provided in this text place within reach organizational concepts normally covered in higher-level undergraduate and graduate courses.

Instructor's Support Materials

This text is accompanied by four diskettes—an educational version of dBASE III PLUS (two diskettes), the Horatio & Co. Cost Control System diskette, and the Horatio & Co. Credit Union System diskette—plus an *Instructor's Manual*.

Both applications are self-contained, menu-driven systems. The cost control system runs under dBASE; the credit union system programs have been compiled, so they do not need dBASE to run. Both systems run in a dual-floppy or hard disk environment. The cost control system needs 384K; the credit union system runs in 256K RAM.

No knowledge of dBASE is assumed, and if you choose, no class time need be spent on the dBASE environment. The discussion of ad hoc reporting in the text is generic and can be applied to any interactive database environment. If you prefer, the dBASE Assistant used as an example of a graphic user interface can be replaced with any other graphic environment. If you want to do more with dBASE, the *Additional Materials* section in Chapter 9 of the *Instructor's Manual* contains material that I have found useful.

I began teaching systems analysis and design in 1979. By 1982, I had abandoned the course in frustration. In 1986, I returned to the course, armed with a set of dBASE exercises that eventually grew into the Horatio & Co. systems. All of the material in the text and *Instructor's Manual* has been refined through classroom testing.

The *Instructor's Manual* is an integral part of the course delivery package. The main text was written for the students; its purpose is to provide concrete experience and to guide observation and conceptualization activities at the students' level. In the *Instructor's Manual*, however, we talk about broader contexts, supplementary issues, motivation, and evaluation.

Each chapter of the *Instructor's Manual* corresponds to one chapter in the text. Each *Instructor's Manual* chapter is composed of nine sections:

   I. Chapter Objectives
   II. Glossary Terms Introduced in This Chapter (170 in all)
   III. Teaching Tips and Strategies
   IV. Transparency Masters
   V. Topic Outline
   VI. Additional Materials
   VII. Answers and Guidelines for Thought Questions (78 in all)
   VIII. Answers and Guidelines for Exercises (94 in all)
   IX. Test Items (30 per chapter—510 in all)

Sections III, VI, VII, and VIII deserve special mention because they support the unique approach used in the text. Because of its concrete and specific nature, students can master the text material on their own. You will find that only a portion of class time need be devoted to discussing the material the students read in the text. The remainder can be spent expanding observations, making more general conceptualizations, and experimenting with the newly learned concepts.

Part III of each chapter presents a plan for carrying out these activities. Part VII provides complete answers to each thought question along with guidelines on using the question to observe and conceptualize. Part VI provides additional relevant materials not presented in the text that can be used to broaden observations and generalize concepts as needed. This section contains *complete* treatments, not just citations and additional references. Finally, Part VIII provides *complete* answers to each exercise along with guidelines on using the exercise to test ideas and generate new experiences.

If you are new to systems courses, the package of text, software, and instructor's manual provides a comprehensive set of materials to guide you through your first offering. As you experience, observe, conceptualize, and experiment with the course, you will tailor the materials to suit your own environment, interests, and style.

If you are an experienced systems teacher, you probably have built your own collection of techniques and materials. Since the topics covered in the text are all "mainstream," substitution and experimentation should be quite simple. Good luck!

Acknowledgments

Many individuals contributed their expertise to the development of this text. I am especially indebted to Karen Calkins of Bryant College; Karen-Ann Kievit of California State University, Long Beach; Anne McClanahan of Ohio University; and John Stoob of Humboldt State University. They need only compare their reviews of earlier drafts with the final product to determine the extent of their contribution.

Special thanks must go to Lisa Gilmour of Ashton-Tate and to Jim Porzak of Bytel Corporation for their vision and enthusiasm and to the trustees of Rider College for granting a research leave to complete this project.

I also wish to thank the supporters and developers of the 1987 AACSB Advanced MIS Faculty Development Institute at Indiana University, especially those involved in the systems area: Ken Kozar, Milt Jenkins, Bob Bostrom, and Nick Vitalari.

Thanks go also to the following people who provided reviews, valuable comments, and other assistance:

Brother William Batt
Manhattan College

Cynthia Mathis Beath
University of Minnesota

Paul Borkowski
Cuyahoga Community College

Patrick Fenton
West Valley College

C. Brian Honess
University of South Carolina

Robert Keim
Arizona State University

Constance Knapp
Pace University

David Kroenke

Richard Leifer
Rensselaer Polytechnic Institute

Charles MaWhinney
Bentley College

Richard Meyer
Hartnell College

R. J. Planisek
Grand Valley State College

Steven Seilheimer
Niagara University

Charles Silcox
Widener University

Ronald Thorn
Grambling State University

James Walters
Pikes Peak Community College

Ronald Williams
Central Piedmont Community College

James Wilson
Triton College

Participants, 1988 National Computer Educators Institute James Madison University

Friends and colleagues at Mitchell Publishing, Inc.

In every team project, there is one person at the center who sets the tone, leads by example, and "makes it all happen." This project is no exception. Erika Berg, editor extraordinaire, did all those things and more. I am truly grateful for the opportunity to know her and to work with her on this project.

At this point, many authors thank their spouses for tolerating long periods of separation and solitude. Fortunately, I do not have to do this. My wife, Camille, compiled detailed class notes from my analysis and design course in 1986 and again in 1987. She read and reread the manuscript, often discovering gaping holes in the development. She provided a second pair of eyes and ears during the feedback process. She kept me laughing. . . . Thanks, Camille.

                                                                                      Bill Amadio

SYSTEMS DEVELOPMENT
A Practical Approach

PART ONE

The Foundation of This Text

Part One of this text contains three chapters. Chapter 1 presents an overview of systems development practice and introduces the systems development life cycle. Chapters 2 and 3 explain how to use the two systems contained on the diskettes accompanying this text. The systems are used at an imaginary organization known as Horatio & Co.

Each of the Horatio systems represents the end product of a systems development project. Subsequent parts of this book will look back in time to reconstruct the procedures and decisions of these projects. In the process, a broad framework of systems development concepts will be developed. The chapters of Part One provide the foundation of this work.

1

CHAPTER 1

Systems Development Practice: An Overview

CHAPTER 2

The Horatio & Co. Cost Control System: Inquiries and Reports

CHAPTER 3

The Horatio & Co. Cost Control System: Maintenance Processes

CHAPTER 1

Systems Development Practice: An Overview

Objectives

Current research in systems analysis and design [8, 9] suggests that a systems analyst should demonstrate

logical ability	mature judgment
thoroughness	practicality
ability to observe	ability to work with others
resourcefulness	dislike of inefficiency
imagination	initiative
oral ability	integrity
abstract reasoning	intelligence
emotional balance	interest in technology
interest in analysis	interest in staff work
writing ability	numerical ability
curiosity	open-mindedness
decisiveness	selling ability
empathy	intuition

and be well versed in

organizational theory	the art of expression
law	information analysis
the art of interviewing	software engineering
project management	programming
economics	databases
user training	

Apparently, the education of such a person is going to take time, possibly a lifetime. Possibly even more than a lifetime [9].

Still with us? Congratulations! Not everyone is capable of making a commitment of "possibly more than a lifetime." This is a good first sign.

In this text the term **systems analysis and design** refers to the activities undertaken by members of a business organization in the development and maintenance of computer-based information systems. This introductory chapter examines this definition more closely and indicates which topics from the rather formidable list on the preceding page are covered in the text.

The specific objectives of this chapter are

1. To identify which members of a business organization participate in the development and maintenance of computer-based information systems;
2. To identify the kinds of computer-based information systems used in business today;
3. To demonstrate how the systems development process is organized;
4. To list the activities undertaken in the development and maintenance of business computer systems;
5. To identify what is covered in this text.

Members of the Systems Development Team

Who participates in the development and maintenance of business computer systems? **Computer programmers** certainly do. They write the instructions that direct the computer to perform its tasks. At this point in your education you probably have had one or more courses in computer programming. The material of this text builds upon that foundation.

The people who use computer systems also participate in the development and maintenance process. In addition, the managers of these people have a stake in the systems their subordinates use. In this text the users and their managers will be referred to as the **user/management group**.

Traditionally **systems analysts** and **systems designers** served as communication links between the programmers and the user/management group. Analysts helped users specify their information requirements. Designers turned those requirements into technical specifications, and programmers turned the technical specifications into working programs for the user/management group. The process is illustrated in Figure 1.1.

When different people or groups of people fill the roles depicted in Figure 1.1, problems abound. Meanings get lost in the translation process from user to analyst to designer to programmer. Natural conflicts develop

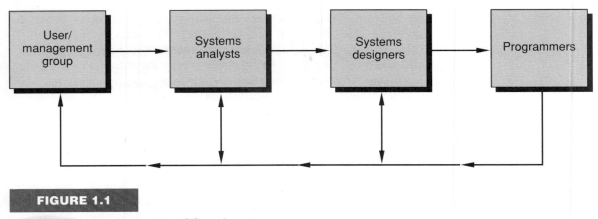

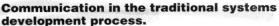

FIGURE 1.1

Communication in the traditional systems development process.

when users, analysts, and programmers must work together but report to different managers. Finally, a system could become obsolete before it is used if business conditions change during the time it takes analysts, designers, and programmers to transform user specifications into a working system of programs.

In response to these problems the roles depicted in Figure 1.1 have collapsed. Today some users know enough about computer systems to serve as their own analyst, designer, and programmer. Today's programmers and analysts are expected to develop business skills as well as technical skills, and in many companies, analysts and programmers now report directly to user management instead of a centralized information systems department. In the future, role boundaries will continue to blur, and users and systems specialists will continue to expand their knowledge of each other's fields. Choices regarding systems development roles will vary depending upon the system and the available development resources.

Kinds of Business Computer Systems

What kinds of computer systems are used in business today? During the 1950s and 1960s there was one answer to that question: accounting. Business managers used computers to process accounts receivable, accounts payable, payroll, and inventory transactions. The objective of these early **electronic data processing (EDP)** applications was to increase transaction processing efficiency by lowering transaction processing costs.

During the early 1970s EDP attention expanded beyond transaction processing efficiency to managerial effectiveness. Managers looked to the organizing power of the computer to transform vast amounts of accounting data into useful information. The terms **information system (IS)** and **management information system (MIS)** were coined to represent the expansion of EDP beyond transaction processing to systems that produce fixed-format inquiries and reports for applications such as sales analysis, financial statements, and cost accounting.

The introduction of the first word processing equipment in the early 1970s marked the start of the **office automation (OA)** movement. These early systems relied upon text editing to provide enhanced productivity to clerical and secretarial workers. Office automation applications have grown to include various forms of electronic mail, electronic conferencing, and image storage, retrieval, and transmission. OA has become an equal partner with IS in business organizations, and the boundaries between the two fields have blurred.

During the late 1970s the minicomputer brought IS power to smaller businesses that could not afford the early mainframe machines. These companies followed the earlier data processing software development pattern. They began with accounting systems and grew into the MIS reporting applications.

Larger companies used minicomputers to decentralize their IS operations. Remote warehouses, factories, and sales offices processed their own data on their minicomputers and sent the results to a central computer. This led to sophisticated applications in inventory control, purchasing, and production control that combined both processing efficiency and managerial effectiveness.

Software tools developed in the late 1970s helped bring about the **decision support systems (DSS)** movement. MIS reports worked well in repetitive situations involving fixed-format historical data, but managers also faced one-of-a-kind problems involving planning for the future. Decision support systems are meant to work in this environment. They depend upon database management systems with report writers and query languages for easy retrieval of unique combinations of data. DSS also rely upon specialized analysis software packages for forecasts and projections. The term **user-friendly software** was coined to describe these software tools. The introduction of this software and the more personal, less repetitive DSS began the collapse of the systems development roles described earlier.

The early 1980s brought a development whose full impact no one could foresee. When IBM introduced its personal computer (PC) in 1981, the company estimated 250,000 units would be sold for all time [1]. By the end of 1987, nearly 15 million PCs and PC-compatibles had been sold. In a few years the popular conception of a computer changed from the monolithic monster in the glass-walled room to the trim three-piece desktop configuration.

Initially, PC users found enhanced productivity in word processing, spreadsheet, file management, and graphics software packages. These personal productivity applications formed the core of the **end-user computing (EUC)** movement. With EUC applications one finds all of the systems development roles collapsed into the user/management group. Some researchers predict that by 1990, 75% of all computer processing capacity will be devoted to EUC applications [2].

The development of early EUC applications led to the need to communicate. For example, downloading data from a mainframe database to a spreadsheet application was easier and more accurate than typing it. Sending a report draft to reviewers and "discussing" it electronically was more efficient than mailing photocopies and arranging a meeting of interested parties. Research and development in **telecommunications** within a multivendor environment of mainframes, minicomputers, office automation systems, and workstations had been going on for quite some time. With millions of new personal computer users on the scene, efforts and competition intensified.

Developments in **artificial intelligence (AI)** and **expert systems** also marked the decade of the 1980s; research focused on the storage of knowledge instead of data. To store knowledge, one needs to represent data and the rules to use the data to solve problems.

One commercially available expert system [7] incorporates the problem-solving expertise of senior financial officers from major corporations and gives advice on products, projects, mergers, and acquisitions. Human experts use the system to examine more alternatives in less time. Nonexperts use the system as a consultant to implement a more sophisticated analysis than would otherwise be possible.

In response to stiff foreign competition in manufacturing, **factory automation** systems began to emerge in the 1980s. Early applications included robots to replace people in difficult and dangerous jobs, computer-aided design (CAD), and computer-integrated manufacturing (CIM) systems. The integration of factory automation, information systems, office automation, and telecommunications technologies is one of the key issues facing information systems management as we approach 1990 [3].

Figure 1.2 summarizes the business computer systems discussed above.

TECHNOLOGY	1960s	1970s			1980s	1990s
Information Systems (IS)	Transaction processing systems	Fixed-format reporting systems	Mini-computers	Decision support systems	End-user computing	Expert systems
Office Automation (OA)		Word-processing equipment		Electronic mail	Teleconferencing	Image processing
Telecommunications			Decentralization of EDP		Local area networking	Enterprise-wide networking
Factory Automation					Computer-integrated manufacturing	

FIGURE 1.2

Business computer systems.

Organizing the Systems Development Process

The preceding discussion of business computer systems shows that any organization can take advantage of computer technology in a number of ways. Even within a small organization one is likely to find systems such as telecommunications and payroll that affect the entire company, systems such as purchasing or salesperson performance analysis that are used at the departmental level, and systems for managing personal files that are developed and maintained by individuals.

If a system is likely to affect the entire company, then the analysis, design, and programming roles should involve systems specialists from the information systems department [4]. This guarantees that new systems conform to company standards and do not disturb existing systems.

In many companies the demand for information systems department resources exceeds the supply. These resources are often controlled by a committee of senior company managers and managers from various departments including information systems. Members of this committee, often called a **steering committee**, review proposals for new information systems and commit resources according to the importance of the proposed system to the company's business objectives. Because of the high demand it is possible for a user/management group to wait several years for information systems department resources.

If a system affects only a few individuals, then systems specialists from the information systems department may not be involved at all or may be involved in an advisory capacity only. In this case the user/management group must provide the expertise to fill the analysis, design, and programming roles. Many companies establish **information centers** to help users develop their own systems. Staffed by systems specialists, these centers provide advice and support for hardware, software, and data questions. Eventually, competent systems specialists will emerge from the user/management groups, and the need for these centers will disappear.

Departmental level systems exhibit the widest range of development teams. Reporting systems such as sales analysis might be developed by the user, whereas a sales forecasting system that affects the budgeting database would require systems specialists to ensure proper control of the process.

Systems Development Activities

Most companies organize development and maintenance activities into a set of phases, commonly called the **systems development life cycle**. Development teams use the life cycle as a blueprint for their activities. Each phase of the life cycle is made up of procedures and decisions. The study of these procedures and decisions comprises the core of this text.

In the early phases of the life cycle the activities are broad and general. Later on, the activities become specific and detailed. Although each organization implements its own version of the life cycle, there are common characteristics that all implementations share. The version of the systems development life cycle used throughout the remainder of this text is illustrated in Figure 1.3.

Problem and/or Opportunity Analysis Phase

The procedures of the **problem and/or opportunity analysis phase** lead to a decision regarding the usefulness of an information system for the problem and/or opportunity at hand. At this point in the life cycle it is too early

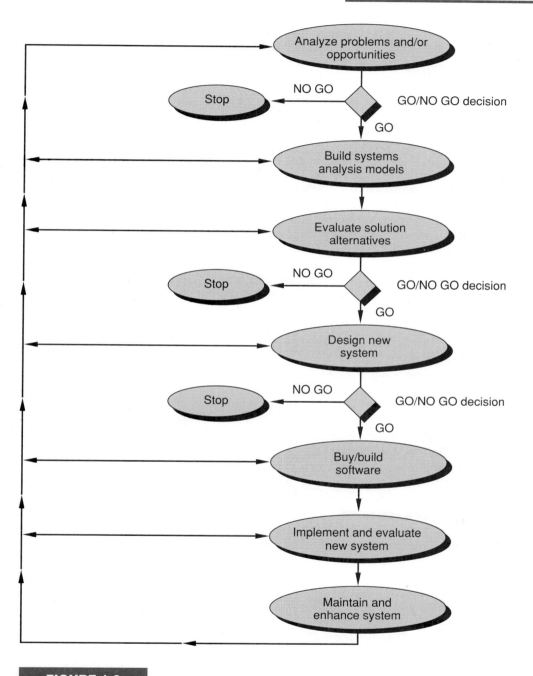

FIGURE 1.3

Systems development life cycle.

to commit to the development of an information system. A favorable decision at the end of this first phase indicates that the development team feels that an information system solution deserves further study. It represents a commitment to the next two phases of the life cycle.

Systems Analysis Model Building Phase

In the **systems analysis model building phase** of the life cycle the development team studies the current means of addressing the problem and/or opportunity at hand. The models serve as a convenient way of describing the present situation and communicating this understanding among development team members.

Once the current situation is modeled, the development team turns its attention toward improving it. The improvements specified at this time become the focus of the remainder of the life cycle. To illustrate their importance, the improvements are commonly referred to as requirements.

Evaluation of Alternatives Phase

In the third phase of the life cycle, **evaluation of solution alternatives**, the development team looks for ways to implement the requirements through an information system. Rough estimates of cost are used to decide how many of the requirements will be addressed and the means by which they will be addressed. In doing this, the team considers the five components of a computer-based information system: hardware, software, data, procedures, and personnel [6].

The development team also evaluates how the new system will be designed and implemented. A decision is made to proceed with the remaining steps of the life cycle or to abandon an information systems solution in favor of some other way of addressing the problem and/or opportunity at hand.

Design Phase and Beyond

If a decision to go ahead is made, the remaining steps of the life cycle vary depending upon the software design and implementation methodology chosen during the evaluation phase. In the **design new system phase**, procedures for the new system are designed, operational and management responsibilities are assigned to individuals, and the physical layout for the hardware is determined.

The variation arises in connection with the data and software components. If the new system is addressing a common problem, such as accounts receivable processing, then software packages are probably already available to perform the functions [4]. In this case software and data design activities are minimal. Instead the development team shops for a package that provides an acceptable match with the requirements.

If the problem and/or opportunity at hand is unique, then the software component of the new system is developed instead of purchased. The software design method depends upon the environment available for the development of the system.

Modern software development environments often contain a **fourth generation language**. Such language provides the capabilities of a relational database management system with such aides to programmer productivity as screen painters, report writers, and nonprocedural programming. Procedural languages such as COBOL, BASIC, and PL/1 fall under the general heading of **third generation languages**.

In addition to fourth generation languages, modern software development organizations are making use of **computer-aided software engineering (CASE)** tools. CASE tools that automate such systems development tasks as program coding and documentation management exist today. Tools that support the more creative tasks of problem analysis, system design, and project management are emerging. Efforts to integrate stand-alone products into a comprehensive environment that provides automated support for the entire life cycle are also underway.

If a fourth generation and/or CASE environment is feasible for a project, then a **prototyping** software design approach is desirable. Prototyping exploits modern technology to produce a working model of a system quickly [5]. The user/management group works with the model or prototype to determine what the final system should do to meet their needs. They communicate their discoveries to the builders who incorporate this feedback into a new version of the prototype. The refinement process is repeated until a satisfactory system is produced.

Prototyping requires a significant contribution from the user/management group. In the prototyping process the user/management group designs the system, and the systems professionals build it [5]. The design function requires a user/management commitment not found in more traditional development methods. This commitment is as necessary as the fourth generation environment for the success of a prototyping project.

If prototyping is not feasible for the new system, then a detailed design of the data and software components of the system is done before any programming begins. This is due to the difficulty of programming systems in third generation languages such as COBOL. Such work requires large commitments from highly skilled and well-trained information systems professionals. While prototyping requires the participation of these same highly skilled professionals, the extent of the commitment is much less.

When the detailed design is complete, the programming costs can be estimated, and the feasibility of the entire project is reviewed. If a decision to proceed is made, the detailed design is "frozen" and programming begins. When all programs have been written and tested, the system is delivered to the users. Because of the difficulty of third generation programming, changes to the original detailed design are usually minimized.

Whatever data and software design method is chosen, the development team assembles an estimate of the costs to develop and maintain the proposed system at the end of the design phase. If a decision to implement the design is made, the software and data components are built or bought in the **buy/build software phase** and combined with the hardware, procedures, and personnel components into a final implementation. A new system is born.

Final Phases of the Life Cycle

During the **implement and evaluate new system phase** of the life cycle the new system is put into production; it performs the work for which it was designed and built. It lives.

Sometimes flaws in the design and programming are found and corrected. Sometimes the environment changes, and the system must be changed to stay current. During the **maintain and enhance system phase** of the life cycle, users and systems specialists work together to maintain the system in a manner similar to that which we use to maintain our bodies.

Inevitably, changes in the environment and advances in technology push system requirements to the point where the current system cannot perform the required work correctly and efficiently. In this case the current system is retired, and the life cycle is complete. If the current system needs to be replaced, the development process begins anew.

What This Text Covers

This text is designed to provide a first experience in systems analysis and design. One system, the Horatio & Co. Cost Control System, runs throughout the text. According to Figure 1.2, the cost control system is an IS application. More specifically, it is a fixed-format reporting system with transaction processing and decision support system components. The purpose of the cost control system is to improve managerial effectiveness through a comprehensive set of reports and inquiries. Its development, using prototyping, will be traced throughout the text.

Chapters 2 and 3 introduce a prototype version of the cost control system; this prototype is contained on one of the diskettes supplied with this text. You will learn to use the prototype as if you were a member of the user/management group. Subsequent chapters look back in time to study the development of the system to this point. The study of the cost control system concludes with a look ahead to the final operational system.

The assignment section of each chapter deals with a second system, the Horatio & Co. Credit Union System, which is also contained on the diskettes supplied with the text. You will do the analysis and make the development decisions for this system as part of your out-of-class assignments.

According to Figure 1.2 the credit union system is also an IS application. The purpose of the credit union system, however, is to improve transaction processing efficiency. The credit union system would be placed to the left of the cost control system on the spectrum of IS applications presented in Figure 1.2. Together both systems illustrate the full range of IS applications from transaction processing through decision support.

The final section of the text presents a general discussion of systems development practice. These chapters build upon your experience with the development of the Horatio & Co. IS applications. The material involves other technologies such as office automation and telecommunications, more complicated organizations than Horatio & Co., and a wider range of systems development tools and techniques than those used for the Horatio systems.

The insights gained from the final section should help you decide what step you will take next as you continue your professional development. Remember, the education of a systems analyst could take "possibly more than a lifetime." Good luck!

References

1. Bender, E. "The Master Plan," *PC World*, Volume 5, Number 8, August 1987, pp. 174–185.

2. Benjamin, R.I. "Information Technology in the 1990s: A Long Range Planning Scenario," *MIS Quarterly*, Volume 6, Number 2, June 1982, pp. 11–31.

3. Brancheau, J., and J. Wetherbe. "Key Issues in Information Systems Management," *MIS Quarterly*, Volume 11, Number 1, March 1987, pp. 23–45.

4. Gremillion, L., and P. Pyburn. "Breaking the Systems Development Bottleneck," *Harvard Business Review*, March-April 1983, pp. 130–137.

5. Jenkins, A. M. "Prototyping: A Methodology for the Design and Development of Application Systems," *Spectrum*, Volume 2, Number 2, April 1985, pp. 1–8.

6. Kroenke, D. *Business Computer Systems*. Santa Cruz, CA: Mitchell Publishing Corporation, 1987.

7. Michaelsen, R., and D. Michie. "Prudent Expert Systems Applications Can Provide a Competitive Weapon," *Data Management*, Volume 24, Number 7, July 1986, pp. 30–34.

8. Shemer, I. "Systems Analysis: A Systemic Analysis of a Conceptual Model," *Communications of the ACM*, Volume 30, Number 6, June 1987, pp. 507–512.

9. Weinberg, G. *Rethinking Systems Analysis and Design*. Boston: Little, Brown and Company, 1985.

CHAPTER 2

The Horatio & Co. Cost Control System: Inquiries and Reports

Objectives

Systems Analysis and Design is a difficult course to take in the classroom because mastery of the subject requires a good deal of practical experience. Chapters 2 and 3 are designed to address this difficulty. They deal with the Horatio & Co. Cost Control System, a menu-driven, fixed-format reporting system supplied on diskette with this text. The objective of these chapters is to provide the background experience required for the realistic study of Systems Analysis and Design.

Chapters 2 and 3 introduce the use of the cost control system. Your responsibilities will be those of the user. You will generate reports, maintain data, and perform all other functions necessary for the proper operation of the system. This will give you some idea of what is supposed to come out of a systems development effort. With a clearer understanding of where you are going, discussions of how to get there will be more meaningful.

The specific objectives of this chapter are

1. To provide experience in the use of a modern computer-based information system;
2. To provide insight into the day-to-day activities of computer systems users.

Introduction

Horatio & Co. Construction is engaged in the development and management of commercial real estate. Throughout this course you will study the cost control system used by Horatio's engineering department.

Based upon revenue forecasts, Horatio & Co. management prepare monthly expense budgets for the engineering department using four General Ledger accounts. These are

ACCOUNT	NAME
4100	Engineering
4200	Materials
4300	Equipment
4400	Subcontractors

The engineering department manager, Sam Tilden, uses the cost control system to keep expenses within the budget guidelines. The system also helps Sam charge expenses to current projects or jobs so that company project managers can bill the clients.

For these tasks Sam relies on the General Ledger (G/L) Budget vs. Actual Reports, the Job Budget vs. Actual Reports, and the Job Cost Reports. The study of the cost control system begins with these **reports**.

Current Status: G/L Budget vs. Actual Reports

As each expense is entered into the system, it is charged to one of the four general ledger accounts. The G/L Budget vs. Actual Report accumulates all expenses charged to each account and compares these totals to the budget figures. The report is available for a particular month or for the full year to date.

To print the reports, choose option 4, REPORTS, from the cost control system main **menu**, Figure 2.1. (To get the most out of this chapter, read the material about running the system first. Study the procedures and output shown in the text, and then try it yourself on the computer.) Instructions for starting the cost control system appear below.

1. The Horatio & Co. Cost Control System runs under dBASE III PLUS. A printer is required if one intends to print reports.

2. Since you will probably want to experiment with the systems, make backup copies before beginning so that the database can be easily restored.

3. To load dBASE III PLUS on dual floppy systems, place the dBASE III PLUS disk #1 in drive A and type DBASE at the DOS A> prompt. When prompted, remove dBASE III PLUS disk #1 from drive A, insert dBASE III PLUS disk #2 in drive A, and press the Enter key.

    If you are using a hard disk system, copy the files on the dBASE III PLUS diskettes to a hard disk directory reserved for dBASE III PLUS.

    To load dBASE III PLUS on hard disk systems, change to the directory reserved for dBASE III PLUS and type DBASE at the DOS prompt.

    If the dBASE III PLUS menu-driven ASSISTANT comes up on program load, hit Esc to exit to the dBASE dot prompt.

4. To run the Horatio & Co. Cost Control System, load dBASE III PLUS, and place the cost control system diskette in your computer's data disk drive. The data disk drive is usually drive B for dual floppy systems and drive A for hard disk systems.

    Dual floppy users should now type the dBASE III PLUS command SET DEFAULT TO B at the dBASE III PLUS dot prompt. When the dot prompt returns, type DO COST. The system is menu-driven from this point on.

```
HORATIO & CO.
COST CONTROL SYSTEM
MAIN MENU

                        1. MAINTAIN BUDGETS

                        2. MAINTAIN EXPENSES

                        3. INQUIRIES

                        4. REPORTS

                        5. ENGINEERS' ACTIVITY ANALYSIS

                        6. PURGE JOBS

                        Q. QUIT

                        Enter choice   Q
```

FIGURE 2.1

Cost control system main menu.

Hard disk users should type the dBASE III PLUS command SET
DEFAULT TO A at the dBASE III PLUS dot prompt. When the dot
prompt returns, type DO COST. The system is menu-driven from this
point on.

5. A word about data entry. When prompting for data, the system displays a highlight bar indicating the maximum length of the field requested. If the entry is shorter than the maximum, press the Enter key
to complete it. If the entry is equal to the maximum, the entry is complete without pressing the Enter key.

6. When entering dates, two slashes will be displayed in the prompt. It is
not necessary to type the slashes with the entry. When entering numeric data with a decimal point, a decimal point will be displayed in
the prompt. Unless the length of the entry equals the maximum, the
user should type the decimal point as part of the entry.

7. Choose option Q, QUIT, from the cost control system's main menu to
return to the dBASE III PLUS dot prompt. Type QUIT to exit dBASE III
PLUS and return to DOS.

The Monthly G/L Report is option 1 on the REPORTS menu, Figure 2.2,
and the Year to Date G/L Report is option 2.

```
HORATIO & CO.
COST CONTROL SYSTEM
REPORTS MENU


                               1. G/L BUDGET VS ACTUAL MONTHLY
                               2. G/L BUDGET VS ACTUAL YEAR TO DATE

                               3. JOB BUDGET VS ACTUAL MONTHLY
                               4. JOB BUDGET VS ACTUAL PROJECT TO DATE

                               5. JOB COST REPORT MONTHLY
                               6. JOB COST REPORT PROJECT TO DATE

                               7. EXPENSE HISTORY REPORT BY ACCOUNT
                               8. EXPENSE HISTORY REPORT BY JOB

                               Q. QUIT

                               Enter choice   Q
```

FIGURE 2.2

PRINTED REPORTS menu.

Each report prompts the user for a cutoff date. No expenses beyond the cutoff date are included in either report. Let's look at examples of these reports that were run with a cutoff date of 02/28/89.

Figure 2.3 shows that for the period 02/01/89 through 02/28/89 total expenses were $28,300, which is within the budget figure of $29,500. However, there were overruns in the engineering line and the subcontractors line. The underpayments in the equipment and materials lines were caused by special discounts of which Horatio was able to take advantage.

The Year to Date G/L Report, Figure 2.4, shows that total expenses for the period 01/01/89 through 02/28/89 are within budget, but individual lines show underpayments and overpayments. Since the variance in the subcontractors line equals the variance from the monthly report, Figure 2.3, this line was exactly on budget at the start of February. The variances in the engineering line show that the department entered February with an overrun of $1200 and added $300 more during the month (verify these figures). Sam Tilden is responsible for investigating the cause of these overruns.

Since the year to date variances for the materials and equipment lines are larger than the corresponding monthly variances, the special discounts had to be in effect before the start of the month. If the discounts are likely to

```
DATE: XX/XX/XX                                                    PAGE:    1
TIME:    XX:XX


                              HORATIO & CO.
                           COST CONTROL SYSTEM
                      G/L BUDGET VS ACTUAL REPORT MONTHLY

                            THROUGH: 02/28/89


      ACCOUNT                          BUDGET      ACTUAL     VARIANCE

      4100  ENGINEERING              12000.00    12300.00     -300.00
      4200  MATERIALS                 7500.00     4000.00     3500.00
      4300  EQUIPMENT                 5000.00     4500.00      500.00
      4400  SUBCONTRACTORS            5000.00     7500.00    -2500.00


      Grand total                    29500.00    28300.00     1200.00
```

FIGURE 2.3

G/L Budget vs. Actual Report—Monthly.

```
DATE: XX/XX/XX                                                    PAGE:    1
TIME:    XX:XX
                              HORATIO & CO.
                           COST CONTROL SYSTEM
                 G/L BUDGET VS ACTUAL REPORT YEAR TO DATE

                          THROUGH: 02/28/89


          ACCOUNT                      BUDGET      ACTUAL    VARIANCE

          4100  ENGINEERING          24000.00    25500.00   -1500.00
          4200  MATERIALS            15000.00     7500.00    7500.00
          4300  EQUIPMENT            10000.00     9000.00    1000.00
          4400  SUBCONTRACTORS       10000.00    12500.00   -2500.00

          Grand total               59000.00    54500.00    4500.00
```

FIGURE 2.4

G/L Budget vs. Actual Report—Year to Date.

continue for a long time, Horatio management may want to adjust the budget figures downward to more accurately reflect anticipated expenses.

Current Status: Job Budget vs. Actual Reports

As each expense is entered into the system, it is charged to a job number in addition to a general ledger account number. The Job Budget vs. Actual Report accumulates all expenses *for a given job* charged to each account and compares this total to a budget figure *for the given job*. The report is available for a particular month or for the full project to date.

The monthly report is option 3 on the REPORTS menu, Figure 2.2, and the project to date report is option 4. The examples on the following page, Figures 2.5 and 2.6, were run with a job number of B107 and a cutoff date of 02/28/89.

```
DATE: XX/XX/XX                                              PAGE:    1
TIME:    XX:XX
                            HORATIO & CO.
                         COST CONTROL SYSTEM
                 JOB BUDGET VS ACTUAL REPORT MONTHLY

                    THROUGH: 02/28/89 JOB: B107


        ACCOUNT                       BUDGET      ACTUAL    VARIANCE

        4100 ENGINEERING             2400.00     2400.00       0.00
        4200 MATERIALS               2500.00     1500.00    1000.00
        4300 EQUIPMENT               5000.00     4500.00     500.00
        4400 SUBCONTRACTORS          5000.00     7500.00   -2500.00

        Grand total                14900.00    15900.00   -1000.00
```

FIGURE 2.5

Job Budget vs. Actual Report—Monthly.

```
DATE: XX/XX/XX                                              PAGE:    1
TIME:    XX:XX
                            HORATIO & CO.
                         COST CONTROL SYSTEM
                JOB BUDGET VS ACTUAL REPORT PROJECT TO DATE

                    THROUGH: 02/28/89 JOB: B107

        ACCOUNT                       BUDGET      ACTUAL    VARIANCE
        4100 ENGINEERING             4800.00     4800.00       0.00
        4200 MATERIALS               5000.00     1500.00    3500.00
        4300 EQUIPMENT              10000.00     9000.00    1000.00
        4400 SUBCONTRACTORS         10000.00    12500.00   -2500.00

        Grand total                29800.00    27800.00    2000.00
```

FIGURE 2.6

**Job Budget vs. Actual Report—Project
to Date.**

Current Status: Job Cost Reports

At the end of each month the engineering department must summarize its expenses by job number and prepare a report for the company project managers. The project managers collect these reports and use them to prepare the monthly client invoices. The Job Cost Report for the engineering department is simple. It shows the job number and the total expenses charged to the job for the month.

To print the reports, choose option 4, REPORTS, from the cost control system main menu, Figure 2.1. The monthly report is option 5 on the REPORTS menu, Figure 2.2, and the project to date report is option 6. The monthly report prompts the user for the cutoff date and reports all expenses entered for the month up to and including the cutoff date. The project to date report prompts the user for the cutoff date and reports all expenses up to and including the cutoff date.

Let's look at examples of each of these reports that were run with a cutoff date of 02/28/89.

The Monthly Job Cost Report, Figure 2.7, shows three jobs active in February: A141, B107, and B762. You will learn more about these jobs when Expense History Reports are discussed later. For now notice that the total

```
DATE: XX/XX/XX                                                      PAGE:    1
TIME:    XX:XX
                              HORATIO & CO.
                           COST CONTROL SYSTEM
                          JOB COST REPORT MONTHLY

              THROUGH: 02/28/89 DEPARTMENT: Engineering


                        JOB                  AMOUNT

                        A141                 9000.00
                        B107                15900.00
                        B762                 3400.00


                        Grand total         28300.00
```

FIGURE 2.7

Job Cost Report—Monthly.

```
DATE: XX/XX/XX                                                    PAGE:    1
TIME:    XX:XX
                            HORATIO & CO.
                         COST CONTROL SYSTEM
                   JOB COST REPORT PROJECT TO DATE

           THROUGH: 02/28/89 DEPARTMENT: Engineering


                   JOB                    AMOUNT

                   A141                 19100.00
                   B107                 27800.00
                   B762                  7600.00


                   Grand total          54500.00
```

FIGURE 2.8

Job Cost Report—Project to Date.

expenses on the Monthly Job Cost Report match the total expenses on the Monthly G/L Budget vs. Actual Report, Figure 2.3. Both equal $28,300. Also notice that the expenses on the Monthly Job Cost Report for Job B107 match the total expenses on the Monthly Job Budget vs. Actual Report, Figure 2.5. Both equal $15,900.

The Project to Date Job Cost Report, Figure 2.8, also shows A141, B107, and B762 as the only active jobs. No jobs have been completed this year. If a job had been completed, then it would appear on the Project to Date Job Cost Report, Figure 2.8, but not on the Monthly Job Cost Report, Figure 2.7. Notice that the expenses on the Project to Date Job Cost Report for Job B107 match the total expenses on the Project to Date Job Budget vs. Actual Report, Figure 2.6. Both equal $27,800.

A Look Back: Expense History Reports

The Budget vs. Actual Reports and the Job Cost Reports present summarized expense information to Sam Tilden. Summarization is one of the most important functions of a computer-based information system. You will see that computers are ideally suited to this type of work.

There are times, however, when a manager needs to see the detail behind the summarized information; year-end audits and job close-outs are examples. For this information Sam turns to the Expense History Reports.

Expense history can be presented by G/L account number or by job number. To print the reports, choose option 4, REPORTS, from the cost control system main menu, Figure 2.1. The account report is option 7 on the REPORTS menu, Figure 2.2, and the job report is option 8. Let's look at examples of each of these reports.

From the G/L Budget vs. Actual Report, Figure 2.3, you know that Horatio & Co. spends approximately $13,000 per month on engineering, account number 4100. The Expense History Report by Account, Figure 2.9, shows the distribution of this amount.

Bob Jones is senior design engineer. He has been with the company for 16 years. Currently, Bob is working on Jobs A141 and B762, which are office park construction projects, Horatio's specialty.

Sam Tilden, the department manager, is working on Job B107, which involves the construction of an air cleaning system for a factory built by Horatio & Co. two years ago. Since Horatio & Co. did not have air cleaning systems engineers on staff, the job was subcontracted to Aspen Engineering, specialists in this field. Sam serves as Horatio contact person for this project, and devotes the rest of his time to the administration of the engineering department.

Sarah Ludwig was recently hired as an engineer. She is working with Bob Jones on Jobs A141 and B762. On Job A141 Sarah is responsible for structural testing. This phase of A141 is coming to a close. As Sarah is phased out of A141, she will take primary responsibility for the design work on Job B762. Notice how the distribution of her time changes from January to February to reflect her changing responsibilities. Notice, also, the change in the distribution of Bob's time. This kind of analysis is typical of the way managers use detailed reports such as the Expense History.

(Do you notice a difference between the presentation of January expenses and February expenses for account 4100? Which one is easier to use? Why?)

The Expense History Report by Account, Figure 2.9, shows expenses for materials, account number 4200, charged to Jobs A141 and B107. Job B762 is in its early stages and has not had materials expense yet.

In Account 4300, EQUIPMENT, notice expenses for the lease of a computer-aided design (CAD) system. This is being used at Aspen Engineering on the B107 project. The $4,500 figure was negotiated at the start of the lease and will remain constant throughout the project.

Account 4400, SUBCONTRACTORS, is of particular interest to us. Recall that this account had an overrun in the month of February. The account is budgeted for $5,000 per month, which represents the agreed upon fee for Aspen Engineering's services.

```
DATE: XX/XX/XX                                              PAGE:      1
TIME:    XX:XX
                            HORATIO & CO.
                         COST CONTROL SYSTEM
                    EXPENSE HISTORY REPORT BY ACCOUNT


DATE        SOURCE           DESCRIPTION            JOB     AMOUNT


*  ACCOUNT: 4100-ENGINEERING

** January
01/25/89 BOB JONES           DESIGN                 A141    3600.00
01/25/89 BOB JONES           DESIGN                 B762    3600.00
01/25/89 SARAH LUDWIG        STRUCTURAL TESTING      A141    3000.00
01/25/89 SARAH LUDWIG        DESIGN                  B762     600.00
01/25/89 SAM TILDEN          CONSULTATION            B107    2400.00
** Subtotal for month                                      13200.00


** February
02/25/89 BOB JONES           DESIGN                  A141    4500.00
02/25/89 SARAH LUDWIG        STRUCTURAL TESTING      A141    2000.00
02/25/89 SAM TILDEN          CONSULTATION            B107    2400.00
02/25/89 BOB JONES           DESIGN                  B762    1800.00
02/25/89 SARAH LUDWIG        DESIGN                  B762    1600.00
** Subtotal for month                                      12300.00


*  Subtotal for account                                    25500.00


*  ACCOUNT: 4200-MATERIALS

** January
01/13/89 NAL-TECH            TEST MATERIALS          A141    3500.00
** Subtotal for month                                       3500.00


** February
02/01/89 ADAMS SUPPLY        PROTOTYPE MATERIALS     B107    1500.00
02/10/89 NAL-TECH            TEST MATERIALS          A141    2500.00
** Subtotal for month                                       4000.00


*  Subtotal for account                                     7500.00
```

FIGURE 2.9

**Expense History Report by Account
(page 1 of 2).**

```
* ACCOUNT: 4300-EQUIPMENT

** January
01/01/89 ETW LEASING          CAD COMPUTER SYSTEM  B107   4500.00
** Subtotal for month                                    4500.00

** February
02/01/89 ETW LEASING          CAD COMPUTER SYSTEM  B107   4500.00
** Subtotal for month                                    4500.00

* Subtotal for account                                   9000.00
* ACCOUNT: 4400-SUBCONTRACTORS

** January
01/01/89 ASPEN ENGINEERING  CLEAN AIR DESIGN       B107   5000.00
** Subtotal for month                                    5000.00

** February
02/01/89 ASPEN ENGINEERING  CLEAN AIR DESIGN       B107   5000.00
02/01/89 ASPEN ENGINEERING  CAD PROGRAMMING        B107   2500.00
** Subtotal for month                                    7500.00

* Subtotal for account                                  12500.00

*** Grand total                                         54500.00
```

FIGURE 2.9

**Expense History Report by Account
(page 2 of 2).**

In February, however, Aspen engineers discovered that the CAD computer software could not do a special set of calculations. A contract programmer had to be hired to supply the necessary software. These expenses were expected, but they could not be accurately estimated in advance. Horatio's contract with the client allows these expenses to be passed along, but there is a $50,000 limit, so Sam will be sure to keep an eye on this situation.

Finally, notice how the grand total of all expenses in the Expense History Report, Figure 2.9, matches the $54,500 total of the Project to Date Job Cost Report, Figure 2.8.

Let's look at an example of the Expense History Report by Job.

The Expense History Report by Job, Figure 2.10, shows the same records you saw in the report by account, Figure 2.9, except the records are

```
DATE: XX/XX/XX                                             PAGE:    1
TIME:    XX:XX
                            HORATIO & CO.
                         COST CONTROL SYSTEM
                     EXPENSE HISTORY REPORT BY JOB


DATE       SOURCE            ACCOUNT            DESCRIPTION            AMOUNT

** JOB A141

* January
01/13/89 NAL-TECH           4200-MATERIALS     TEST MATERIALS         3500.00
01/25/89 BOB JONES          4100-ENGINEERING   DESIGN                 3600.00
01/25/89 SARAH LUDWIG       4100-ENGINEERING   STRUCTURAL TESTING     3000.00
* Subtotal for month                                                 10100.00

* February
02/10/89 NAL-TECH           4200-MATERIALS     TEST MATERIALS         2500.00
02/25/89 BOB JONES          4100-ENGINEERING   DESIGN                 4500.00
02/25/89 SARAH LUDWIG       4100-ENGINEERING   STRUCTURAL TESTING     2000.00
* Subtotal for month                                                  9000.00

** Subtotal for job                                                  19100.00

** JOB B107

* January
01/01/89 ETW LEASING        4300-EQUIPMENT     CAD COMPUTER SYSTEM    4500.00
01/01/89 ASPEN ENGINEERING  4400-SUBCONTRACTORS CLEAN AIR DESIGN      5000.00
01/25/89 SAM TILDEN         4100-ENGINEERING   CONSULTATION           2400.00
* Subtotal for month                                                 11900.00

* February
02/01/89 ETW LEASING        4300-EQUIPMENT     CAD COMPUTER SYSTEM    4500.00
02/01/89 ASPEN ENGINEERING  4400-SUBCONTRACTORS CLEAN AIR DESIGN      5000.00
02/01/89 ASPEN ENGINEERING  4400-SUBCONTRACTORS CAD PROGRAMMING       2500.00
02/01/89 ADAMS SUPPLY       4200-MATERIALS     PROTOTYPE MATERIALS    1500.00
02/25/89 SAM TILDEN         4100-ENGINEERING   CONSULTATION           2400.00
* Subtotal for month                                                 15900.00
** Subtotal for job                                                  27800.00
```

FIGURE 2.10

**Expense History Report by Job
(page 1 of 2).**

```
** JOB B762

* January
01/25/89 BOB JONES          4100-ENGINEERING     DESIGN              3600.00
01/25/89 SARAH LUDWIG       4100-ENGINEERING     DESIGN               600.00
* Subtotal for month                                                 4200.00

* February
02/25/89 BOB JONES          4100-ENGINEERING     DESIGN              1800.00
02/25/89 SARAH LUDWIG       4100-ENGINEERING     DESIGN              1600.00
* Subtotal for month                                                 3400.00

** Subtotal for job                                                  7600.00


*** Grand total                                                     54500.00
```

FIGURE 2.10

**Expense History Report by Job
(page 2 of 2).**

grouped by job in this report. Notice that both grand totals equal $54,500, the figure also shown in the Project to Date Job Cost Report, Figure 2.8. Notice, also, that the February subtotals for each job match the figures reported for each job in the Monthly Job Cost Report, Figure 2.7, and that subtotals for each job in the Expense History Report by Job match the figures reported for each job in the Project to Date Job Cost Report, Figure 2.8.

Sam Tilden reviews this report before submitting the summarized Job Cost Report to the project managers. He makes sure that the totals from the two reports match. He then reviews the detailed Expense History Report by Job, Figure 2.10, for completeness because any expenses not listed on the report will not be billed to the clients. The mechanisms used to control the accuracy and the completeness of data entry will be discussed later.

A Look Back: Inquiries

A full set of printed reports are run at the end of each month. This gives Sam Tilden both a summarized and detailed account of the month's activities and the current status of his department. During the month questions about individual expenses, accounts, and jobs come up. These are usually answered using the **INQUIRIES** function of the cost control system.

To make an inquiry, choose option 3, INQUIRIES, from the cost control system main menu, Figure 2.1. On the INQUIRIES menu, Figure 2.11, IN-QUIRY BY MONTH is option 1, BY SOURCE is option 2, BY ACCOUNT is option 3, BY JOB is option 4, and BY JOB AND ACCOUNT is option 5.

Let's look at an example. Suppose Sam Tilden wanted to study Nal-Tech's charges in isolation. He would do an INQUIRY BY SOURCE and enter NAL as the source value. The information in Figure 2.12 appears on the screen.

This information could have been retrieved from the Expense History Report by Account, Figure 2.9, but the inquiry is used here because the question concerns Nal-Tech only. In the Expense History Report, Nal-Tech's information is intermingled with all of the other suppliers' figures. If Sam used the report, he would have to find all of Nal-Tech's records and organize them by hand. This process is time consuming and prone to error. The ability to select a few records from a large database and present them in isolation is one of the chief advantages of modern computer-based information systems.

```
HORATIO & CO.
COST CONTROL SYSTEM
INQUIRIES MENU


                        1. EXPENSES BY MONTH

                        2. EXPENSES BY SOURCE

                        3. EXPENSES BY ACCOUNT

                        4. EXPENSES BY JOB

                        5. EXPENSES BY JOB AND ACCOUNT

                        Q. QUIT

                        Enter choice   Q
```

FIGURE 2.11

SCREEN INQUIRIES menu.

```
DATE: XX/XX/XX                                              PAGE:    1
TIME:   XX:XX
                        EXPENSES BY SOURCE FOR NAL

SOURCE              ACCOUNT              JOB DESCRIPTION        AMOUNT   HOURS

* January
NAL-TECH            4200 MATERIALS       A141 TEST MATERIALS    3500.00     0
* Subtotal for month                                           3500.00     0

* February
NAL-TECH            4200 MATERIALS       A141 TEST MATERIALS    2500.00     0
* Subtotal for month                                           2500.00     0

** Grand total                                                 6000.00     0
```

FIGURE 2.12

INQUIRY BY SOURCE = NAL.

A second advantage of the inquiry for this type of question involves completeness. The Expense History Report represents the status of the system at the time the report was printed. As soon as one new transaction is entered into the system, the report is obsolete. The inquiry, on the other hand, always presents all records that have been entered to date. It is true that Sam could have rerun the Expense History Report, but then he would be printing all expense records on file just to see two of them.

The remaining inquiry functions work in the same way. They offer quick access and simple presentation to answer questions concerning *individual* expense items, accounts, or jobs. Take some time to experiment with them.

Engineers' Activity Analysis

Construction projects require human resources in addition to financial resources. As engineering department manager, Sam Tilden is also responsible for the control of the department's human resources. When dealing with people instead of dollars, Sam uses the ENGINEERS' ACTIVITY ANALYSIS option of the cost control system. Let's look at an example.

Suppose Sam wants to study Sarah Ludwig's changing responsibilities. To run the analysis, choose option 5, ENGINEERS' ACTIVITY ANALYSIS, from the cost control system main menu, Figure 2.1, and enter LUDWIG as the engineer's name. The information in Figure 2.13 appears on the screen.

It is easy to see that February brought a large increase in Sarah's design responsibility. She jumped from 15 hours to 40 hours worth of design work in February, and her structural testing responsibility decreased by the same 25 hours.

The ENGINEERS' ACTIVITY ANALYSIS also presents a percentage breakdown of Sarah's time. Notice that the figures are rounded to the nearest whole number before they are displayed. This sometimes causes errors in the display of subtotal and grand total calculations. (Why do you think the system was designed this way? If you were Sam Tilden, would you accept this design? Why or why not?)

Bob Jones was Sarah's partner on job A141. Run an ACTIVITY ANALYSIS on him and compare the results to Figure 2.13 to determine the effect of Sarah's increased design responsibility on Bob's work assignments.

```
DATE: XX/XX/XX                                              PAGE:    1
TIME:    XX:XX
                        ENGINEERS' ACTIVITY ANALYSIS
                      ANALYSIS BY ENGINEER FOR LUDWIG

DATE        SOURCE              DESCRIPTION          JOB    HOURS      %

01/25/89    SARAH LUDWIG        DESIGN               B762    15        8
02/25/89    SARAH LUDWIG        DESIGN               B762    40       22

   * Subtotal for DESIGN                                     55       30

01/25/89    SARAH LUDWIG        STRUCTURAL TESTING   A141    75       41
02/25/89    SARAH LUDWIG        STRUCTURAL TESTING   A141    50       27

   * Subtotal for STRUCTURAL TESTING                        125       69

  ** Grand total                                            180      100
```

FIGURE 2.13

ENGINEERS' ACTIVITY ANALYSIS for LUDWIG.

Summary

This chapter presents a look at the output component of the cost control system. Through the use of inquiries and reports, Sam Tilden is able to manage the engineering department effectively.

The next chapter presents a look at the input and processing components of the cost control system. Through these components users maintain the database that supports the output component presented in this chapter.

THOUGHT QUESTIONS
In Your Opinion. . .

1. What is the difference between an inquiry and a report?

2. When running cost control system inquiries and reports, you may have noticed the prompt Printer/Screen/Quit. For what situations is printed output appropriate? For what situations is screen output appropriate? Do you think the answers to these questions depend upon the person who is requesting the inquiry or report? Why or why not?

3. Did you notice the date and time displayed at the top of each inquiry and report? What are the benefits of this design feature?

4. How do you like using a menu-driven system? How does it compare to other computer-based systems you have used? Do you think a menu-driven design is well suited to the work of the cost control system? What kind of work would not be well suited to a menu-driven design?

5. Put yourself in Sam Tilden's position. Which three cost control system inquiries and reports do you think are used most frequently? Why? Which three are used least frequently? Why? Would you change anything about the most frequently used inquiries and reports? Can you think of an inquiry or report that is not included in the Horatio & Co. Cost Control System?

HORATIO&CO.

System Overview

Chapters 2–14 contain sets of exercises that refer to The Horatio & Co. Employee Credit Union System. The credit union provides many financial services to Horatio employees. The system you will study maintains the credit union's savings accounts. Horatio employees make deposits to and withdrawals from these accounts. Interest is credited at the end of each month, and a maintenance fee is charged at the end of the month if the account balance falls below $1000 at any time during the month. The credit union manager is Camille Abelardo (code 2Q3). Walter O'Reilly (code 774) serves as window teller.

Camille and Walter are responsible for the day-to-day operation of the union. This involves recording deposits and withdrawals, maintaining accurate records of the accounts, reconciling cash balances at the end of each day, and preparing reports for members and the Internal Revenue Service.

The system runs on a commercial timesharing network. The union pays a monthly fee, based upon the volume of activity, for the use of the system. Terminals are located at Camille's desk and at the credit union's teller windows.

Instructions for starting the microcomputer version of the credit union system follow.

1. The Horatio & Co. Credit Union System does not require dBASE III PLUS. It runs directly from DOS. A printer is required if one intends to print reports.

2. Since you will probably want to experiment with the systems, make backup copies before beginning so that the database can be restored easily.

3. To run the credit union system, load DOS and place the assignment system diskette in your computer's A drive. At the A> prompt, type CREDIT. The system is menu-driven from this point on.

4. The credit union system menus have an added feature. The up-arrow and down-arrow keys can be used to highlight the desired option. Once the desired option is highlighted, it is activated by pressing the Enter

key. Of course, choices can still be activated by pressing the number corresponding to the desired option.

5. The rules of data entry for the cost control system apply to the credit union system as well.

6. Choose option 7, END, from the credit union system's main menu to return to DOS.

The credit union system main menu, Figure 2.1A, appears below. Notice the similarity to the cost control system main menu.

Screen Inquiries

The SCREEN INQUIRIES menu, Figure 2.2A, is used by both Walter and Camille. Walter uses option 1, ACCOUNT INQUIRY, to answer questions at the teller's window. The inquiry shows the current month's beginning balance, deposits, withdrawals, and current balance. Accounts are identified by a 4-character account number. Help with identifying a member's account number is available by pressing the F1 key at the prompt for the number.

Camille and Walter use separate cash drawers for their teller activities. Camille uses option 2, CASH DRAWERS INQUIRY, to determine how

```
HORATIO & CO. EMPLOYEE CREDIT UNION
------------------------------------

                    1. MAINTAIN ACCOUNTS

                    2. MAINTAIN TRANSACTIONS

                    3. SCREEN INQUIRIES

                    4. PRINTED REPORTS

                    5. CLOSE OUT MONTH

                    6. END OF YEAR

                    7. END
```

FIGURE 2.1A

Credit union system main menu.

```
SCREEN INQUIRIES
----------------

                    1. ACCOUNT INQUIRY

                    2. CASH DRAWERS INQUIRY

                    3. END
```

FIGURE 2.2A

SCREEN INQUIRIES menu.

much cash should be in the drawers at any given time. This serves as a cash control; the amount reported by the inquiry is compared to an actual cash count several times per day. The inquiry also allows Camille to check the cash on hand without disturbing Walter's teller activities. The program is especially valuable when extra tellers are added during busy periods. Camille starts each day with $1500 in cash in each cash drawer.

Printed Reports

There are six options on the PRINTED REPORTS menu, Figure 2.3A.

Option 1, TOTAL ON-DEPOSIT REPORT, shows the name and number of each account holder along with the current balance. Camille prints this report once per month and files it with the government regulatory agencies.

Option 2, MONTHLY STATEMENTS, is similar to ACCOUNT IN-QUIRY of the SCREEN INQUIRIES menu, Figure 2.2A. These are printed once per month and mailed to the account holders. For each account, the report shows the number, name, and address of the account holder, the opening balance for the current month, and all deposits and withdrawals. The program computes the monthly interest using the average daily balance and the interest rate for the current month, which is input by the user at the start of the report. Finally, the program assesses a maintenance fee of $5 if the account balance fell below $1000 at any time during the month.

The DAILY ACTIVITY LOGS are option 3 of the PRINTED REPORTS menu. They are used for the end-of-day cash reconciliation. To run a log, the user enters the code of the teller and the date desired. The log shows the opening cash drawer balance of $1500, and all deposits and withdrawals handled by the teller for the date entered. Cash deposits and withdrawals

```
PRINTED REPORTS
---------------

              1. TOTAL ON-DEPOSIT REPORT

              2. MONTHLY STATEMENTS

              3. DAILY ACTIVITY LOGS

              4. DAILY INTEGRITY REPORT

              5. YEAR END 1099S

              6. END
```

FIGURE 2.3A

PRINTED REPORTS menu.

are identified, and their effect on the cash drawer balance is shown. At the end of the day each teller must run his or her log and certify that the transactions are correct and that the cash drawer balance shown agrees with the cash drawer amount.

The DAILY INTEGRITY REPORT, option 4, is run at the end of each day for additional accuracy control. It shows the totals of the monthly opening balances, deposits and withdrawals prior to the date of the report, deposits and withdrawals on the date of the report, and current balances. More will be said about this report in the later chapters.

YEAR END 1099S, option 5, is run at the end of the year on two-part pre-printed forms. One copy is mailed to the account holder; the other is mailed to the Internal Revenue Service.

Exercises

1. Call up the credit union system main menu. Choose option 4, PRINTED REPORTS, and run option 1, TOTAL ON-DEPOSIT REPORT. Determine how many account holders are on the system. What are their

names, numbers, and balances? Observe the total of all accounts and file your copy of this report for later use.

2. According to the system, the current month is September 1988. Run option 2, MONTHLY STATEMENTS, from the PRINTED REPORTS menu. In response to the prompt to enter the annual percentage rate for this month, type the digits 06 with no decimal point and no Enter key. This means that interest will be calculated at an annual rate of 6% for one month. The interest calculation is INTEREST = AVERAGE DAILY BALANCE x (.06/12). This is standard operating procedure for the banking industry.

   When the statements are printed, notice that only September deposits and withdrawals are shown. The balance forward for each account is the account balance as of September 1, 1988, the beginning of the current month.

   What account holder received the most interest for the month? What accounts were charged maintenance fees? Determine when the balances of these accounts fell below $1000.

   Notice that the account balances listed on the TOTAL ON-DEPOSIT REPORT match the last balances printed on the MONTHLY STATEMENT *before* the calculation of interest and maintenance fee. File your copy of the MONTHLY STATEMENTS for later use.

3. Print the DAILY ACTIVITY LOG for Walter O'Reilly on September 13, 1988. Run option 3, DAILY ACTIVITY LOGS, of the PRINTED REPORTS menu. Enter the characters 774 with no Enter key in response to the prompt to enter the teller's code. Enter the digits 091388 with no slashes and no Enter key in response to the prompt to enter the desired date.

   How many transactions did Walter handle on September 13? How many were deposits? How many were withdrawals? How many were cash transactions? How many were check transactions? Is the cash calculation correct?

4. Run option 4, DAILY INTEGRITY REPORT, using a date of 093088. Verify the integrity of the database using a combination of the reports run in the previous exercises and manual calculations.

5. Run option 5, YEAR END 1099S from the PRINTED REPORTS menu. File your copy of this report.

6. From the SCREEN INQUIRIES menu, run option 1, ACCOUNT INQUIRY. Use the help function to determine the number of the account belonging to Erika Berg and enter the number at the prompt. Notice that the last balance printed matches the current balance reported in the TOTAL ON-DEPOSIT REPORT from exercise 1.

Do an ACCOUNT INQUIRY for each of the other three accounts. Verify the last balance printed with those reported on the TOTAL ON-DEPOSIT REPORT.

7. Run option 2, CASH DRAWERS INQUIRY, from the SCREEN IN-QUIRIES menu. Enter the digits 091388 with no slashes and no Enter key in response to the prompt to enter the desired date.

   For teller 774 does the cash amount match the ending cash balance from the DAILY ACTIVITY LOG printed in exercise 3? Go back to the printed reports menu and run a DAILY ACTIVITY LOG for teller 2Q3 on September 13, 1988. Verify that the cash balance of the CASH DRAWERS INQUIRY matches the DAILY ACTIVITY LOG for teller 2Q3.

CHAPTER 3

The Horatio & Co. Cost Control System: Maintenance Processes

Objectives

The chief characteristic of a fixed-format reporting system is a comprehensive set of predefined reports and inquiries. The previous chapter introduced you to this component of the Horatio & Co. Cost Control System.

The data for information system reports and inquiries is usually maintained by transaction processing modules. In the cost control system, data maintenance functions are provided in options 1. MAINTAIN BUDGETS, 2. MAINTAIN EXPENSES, and 6. PURGE JOBS of the cost control system main menu, Figure 2.1. These modules are discussed in this chapter.

The specific objectives of this chapter are

1. To provide experience with computer-based transaction processing;
2. To mold the experiences with the cost control system into a model to use as a guide in the study of systems analysis and design.

Day-to-day Activity: Expense Maintenance Overview

Like many other IS applications (see Figure 1.2), Horatio & Co.'s control system processes transactions. The database of any **transaction processing system** consists of one or more **master files** supported by any number of **transaction files**. The master files hold background information about the entities represented by the system, while the transaction files are used to record day-to-day activity.

In this text a group of related files is called a **data store**. In the cost control system there is one master data store called BUDGETS. BUDGETS is made up of an identification record file and a budget record file. In the cost control system there is one transaction data store called EXPENSES. EXPENSES is made up of a single expense record file.

Users enter data into transaction data stores. This is part of a process called **transaction maintenance**. The system uses the transaction data to make permanent changes in the master and/or transaction data stores. This process is known as **updating**. The extent of the changes depends upon the design of the system.

Summary reports such as the Budget vs. Actual Reports and detail reports such as the Expense History Report use the master and/or transaction data. The **cycle of transaction maintenance-update-reports** is fundamental to the operation of transaction processing systems. Figure 3.1 summarizes the cycle.

Contents of the Database

At this time BUDGETS, the cost control system master data store, contains twenty-one budget records, one for each of the account/job combinations for which expenses were/will be incurred during the months January, February, and March, 1989. Each record contains the account number, the job number, the month ending date, and the monthly budget for the account/job combination.

The BUDGETS data store also contains four identification records, one for each of the four general ledger accounts. Each record contains the number of the account and the name of the account.

In Figure 3.2 you see a listing of the data stored in BUDGETS. The **sequence** of the budget records is ACCOUNT, JOB, MONTH END-

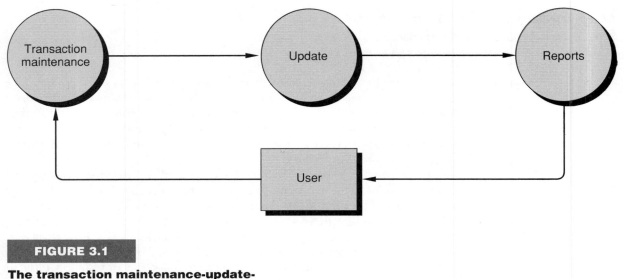

FIGURE 3.1

The transaction maintenance-update-
reports cycle.

ING DATE. The sequence of the identification records is ACCOUNT (verify this).

At present, EXPENSES, the cost control system transaction data store, is made up of a single file containing 18 records. This number will increase as more transactions are entered into the system. Each record contains the date of the expense, the general ledger account to be charged for the expense, the job to be charged for the expense, the source of the expense, a short description of the expense, the amount of the expense, and Horatio employee hours, if any, connected with the expense.

In Figure 3.3 you see a listing of the data stored in EXPENSES displayed in DATE, ACCOUNT, JOB, SOURCE, DESCRIPTION sequence (verify this). The reports display this data in the required sequences automatically.

Running Expense Maintenance

To run transaction maintenance on the EXPENSES file, choose option 2, MAINTAIN EXPENSES, from the cost control system main menu, Figure 2.1.

```
------BUDGET RECORDS------          ---ID RECORDS---
         MONTH     MONTHLY
ACCT JOB END DATE  BUDGET       ACCT NAME

4100 A141 01/31/89  6000.00     4100 ENGINEERING
4100 A141 02/28/89  6000.00     4200 MATERIALS
4100 A141 03/31/89  6000.00     4300 EQUIPMENT
4100 B107 01/31/89  2400.00     4400 SUBCONTRACTORS
4100 B107 02/28/89  2400.00
4100 B107 03/31/89  2400.00
4100 B762 01/31/89  3600.00
4100 B762 02/28/89  3600.00
4100 B762 03/31/89  3600.00
4200 A141 01/31/89  5000.00
4200 A141 02/28/89  5000.00
4200 A141 03/31/89  5000.00
4200 B107 01/31/89  2500.00
4200 B107 02/28/89  2500.00
4200 B107 03/31/89  2500.00
4300 B107 01/31/89  5000.00
4300 B107 02/28/89  5000.00
4300 B107 03/31/89  5000.00
4400 B107 01/31/89  5000.00
4400 B107 02/28/89  5000.00
4400 B107 03/31/89  5000.00
```

FIGURE 3.2

**Listing of records in BUDGETS data
store files.**

On the MAINTAIN EXPENSES screen, Figure 3.4, you see the available
options presented along the bottom of the screen. An option is invoked by
pressing the key corresponding to the first letter of the option. Above the
option menu line the cost control system displays the current EXPENSES
record. At this time the current record is the first record in the sequence
(check this using Figure 3.3).

Begin your work in MAINTAIN EXPENSES by pressing H for Help.
Press any key to move from one help screen to the next. The prompt More...
at the bottom of the screen indicates that other screens are waiting. The
prompt OK... at the bottom of the screen means that the current screen is
the final screen.

DATE	ACCT	JOB	SOURCE	DESCRIPTION	AMOUNT	HOURS
01/01/89	4300	B107	ETW LEASING	CAD COMPUTER SYSTEM	4500.00	0
01/01/89	4400	B107	ASPEN ENGINEERING	CLEAN AIR DESIGN	5000.00	0
01/13/89	4200	A141	NAL-TECH	TEST MATERIALS	3500.00	0
01/25/89	4100	A141	BOB JONES	DESIGN	3600.00	40
01/25/89	4100	A141	SARAH LUDWIG	STRUCTURAL TESTING	3000.00	75
01/25/89	4100	B107	SAM TILDEN	CONSULTATION	2400.00	40
01/25/89	4100	B762	BOB JONES	DESIGN	3600.00	40
01/25/89	4100	B762	SARAH LUDWIG	DESIGN	600.00	15
02/01/89	4200	B107	ADAMS SUPPLY	PROTOTYPE MATERIALS	1500.00	0
02/01/89	4300	B107	ETW LEASING	CAD COMPUTER SYSTEM	4500.00	0
02/01/89	4400	B107	ASPEN ENGINEERING	CAD PROGRAMMING	2500.00	0
02/01/89	4400	B107	ASPEN ENGINEERING	CLEAN AIR DESIGN	5000.00	0
02/10/89	4200	A141	NAL-TECH	TEST MATERIALS	2500.00	0
02/25/89	4100	A141	BOB JONES	DESIGN	4500.00	50
02/25/89	4100	A141	SARAH LUDWIG	STRUCTURAL TESTING	2000.00	50
02/25/89	4100	B107	SAM TILDEN	CONSULTATION	2400.00	40
02/25/89	4100	B762	BOB JONES	DESIGN	1800.00	20
02/25/89	4100	B762	SARAH LUDWIG	DESIGN	1600.00	40

FIGURE 3.3

Listing of EXPENSES data.

Figure 3.5 explains each option in the MAINTAIN EXPENSES option menu. The options Add, Del, and Modify form the foundation of the transaction maintenance module. System users maintain an accurate collection of expenses by adding, deleting, and modifying EXPENSES records. Do not try these options until the next section. If you accidentally choose one of these options, hit the Escape key to cancel the choice and answer No to any questions to save or delete records. Eventually, you will bring the MAINTAIN EXPENSES option menu back to the screen.

The Tally option counts the records. As usual, press any key to continue past the display of the record count.

The List option allows the user to see the EXPENSES records. The list begins with the current record, the one displayed above the option menu, and runs to the end of the file. If a listing of all records is desired, choose the Beg option first to make the first record the current record, and then choose List. Figure 3.3 was created in this way. Try it.

```
HORATIO & CO.
COST CONTROL SYSTEM
MAINTAIN EXPENSES

Date 01/01/89


Account                           Account
Number    4300                    Name        EQUIPMENT


Job  B107


Source   ETW LEASING


Description   CAD COMPUTER SYSTEM


Amount    4500.00                 Hours     0

Ret/Beg/End/Next/Prev/Skip/Modify/Add/Copy/Del/List/Filt/Tally/Help/Quit? M
```

FIGURE 3.4

MAINTAIN EXPENSES screen.

The Beg, End, Next, Prev, Ret, and Skip options are used to get to specific locations in EXPENSES. Beg, End, Next, Prev, and Skip are self-explanatory. The Ret option presents a blank screen and allows the user to fill in specific values for the highlighted fields. When the user accepts the specification screen by pressing PgDn or by making an entry in the last highlighted field, the system retrieves the first record in the sequence that matches the user's specifications. These options are quite safe. Experiment with them. Use Figure 3.3 to verify your results.

The Filt option is interesting. The user fills in specific values in any field, and the system behaves as if only records that match the specifications exist. Try it. Fill in 4100 for the Account Number on the Filter screen, then press PgDn to accept and release the specification. Choose options such as Tally, List, Next, and Prev, and observe the results. When you are finished, choose Filt again and cancel the filter.

The MAINTAIN EXPENSES option menu contains a great deal of information. It can be overwhelming at first sight. Keep practicing until the organization of the menu and the purpose of each option is clear.

Option	Database maintenance action
Add	add a record to the database
Beg	go to beginning of database
Copy	duplicate current record
Del	delete current record
End	go to end of database
Filt	set filter on database
List	display records on screen
Modify	edit current record
Next	go to next record
Prev	go to previous record
Quit	terminate current activity
Ret	retrieve a record by key
Skip	move up or down by a specified number of records
Tally	count records
White key	Screen editing action
$\rightarrow$	character right
$\leftarrow$	character left
up arrow	previous field
down arrow	next field
PgDn, PgUp	accept screen
End	next word/field
Home	previous word/field
Del	delete character
Ins	insert on/off toggle

FIGURE 3.5

Cost control system Help screens.

Entering Expenses

It is time to consider the Add, Del, and Modify options. These options are different from the ones you have used so far because these options change the EXPENSES data. To enter an EXPENSES transaction, press A for Add. Enter the data shown in Figure 3.6. No entry is made in the Account Name field. If you make a mistake before the prompt to save the record, use the

```
HORATIO & CO.
COST CONTROL SYSTEM
MAINTAIN EXPENSES

Date 03/01/89        (DO NOT ENTER THE SLASHES)


Account                              Account
Number     4301                      Name        (NO ENTRY HERE)



Job        B107



Source     ETW LEASING



Description    CAD COMPUTER SYSTEM



Amount     4500.00                   Hours       0

Ret/Beg/End/Next/Prev/Skip/Modify/Add/Copy/Del/List/Filt/Tally/Help/Quit? A
```

FIGURE 3.6

Data for new EXPENSES transaction.

up arrow and down arrow cursor control keys to move the cursor to the
incorrect field and type the correct information.

Once the Hours figure is entered, the system checks the validity of the
entry. The warning NO MATCHING BUDGET RECORD indicates that the
entry is questionable because a monthly budget record has not been established for Account 4301, Job B107 in March 1989. Press Y to display the
account number-job number-month combinations for which budget records
have been established. Press any key at the OK… prompt; press N to refuse
to override the warning, and then press Y to correct the invalid entry.

Use the down arrow key to get to the Account Number field. Change
the entry to 4300, and press the PgDn key to accept and release the updated
screen. This time the system finds a matching budget record and displays EQUIPMENT, the corresponding Account Name. Press Y to save the
new record.

The MAINTAIN EXPENSES program is designed to Add a batch of
records in one sitting, so the program is now waiting to accept your next
entry. Press the Escape key to terminate this Add session.

Choose the Tally option to see the new record added to the EXPENSES count. Choose the Beg option to get to the beginning of the EXPENSES, and then choose the List option to see the new record at the end of the list of EXPENSES. The new record is shown at the end because it is the only transaction for March at this time. Recall that the sequence of EXPENSES is Date, Account, Job, Source, Description.

If you made a mistake, retrieve the new record by pressing R and entering a Date of 03/01/89. No other entries are necessary, so press PgDn to accept and release the screen. Choose the Modify option by pressing M. Use the arrow keys to position the cursor to the incorrect field, and type the correct information. Press PgDn to accept and release the updated record, and answer the save question with Y.

When the transaction is correctly entered, return to the cost control system main menu, Figure 2.1 by pressing Q to choose the Quit option. There will be a slight delay as the system performs the update. The cost control system is designed to perform its update function automatically upon exit from the MAINTAIN EXPENSES program.

To see the results of your work, do an INQUIRY BY MONTH and enter 03/31/89 for the month ending date to see your transaction. An INQUIRY BY ACCOUNT for Account 4300 will also show it. Finally, run a G/L Budget vs. Actual Monthly Report using a cutoff date of 03/31/89 to see the transaction combined with BUDGETS master data.

Once you have entered and verified the transaction, enter the new expense transactions shown in Figure 3.7. The fourth transaction has no matching budget record. Override the warning message when it appears.

DATE	ACCT	JOB	SOURCE	DESCRIPTION	AMOUNT	HOURS
03/01/89	4400	B107	ASPEN ENGINEERING	CLEAN AIR DESIGN	5000.00	0
03/01/89	4200	B107	ADAMS SUPPLY	PROTOTYPE MATERIALS	4000.00	0
03/10/89	4200	A141	NAL-TECH	TEST MATERIALS	1500.00	0
03/15/89	4200	B762	ADAMS SUPPLY	FOUNDATION BLOCKS	2000.00	0
03/25/89	4100	A141	BOB JONES	DESIGN	4950.00	55
03/25/89	4100	B762	BOB JONES	DESIGN	990.00	11
03/25/89	4100	B107	SAM TILDEN	CONSULTATION	2400.00	40
03/25/89	4100	A141	SARAH LUDWIG	STRUCTURAL TESTING	1200.00	30
03/25/89	4100	B762	SARAH LUDWIG	DESIGN	2400.00	60

FIGURE 3.7

Data for March expense transactions.

Run a Monthly Job Budget vs. Actual Report for Job B762 using a cutoff date of 03/31/89 to see the effect of the override. (Why do you think the system was designed to allow expenses for which matching budget records do not exist?)

The details of cost control system data entry are summarized in Figure 3.8. With a little practice these will become second nature to you.

1. When adding or modifying a record, use the up arrow and down arrow keys to move from one field to another.

2. A screen of data is released for processing whenever an entry is made in the last field. Use the PgDn key to accept and release a screen of data from some place other than the last field.

3. When adding new records, terminate the Add session and return to the maintenance option menu by pressing Escape while the system is waiting to accept a new entry.

4. The current record is the one displayed in the layout above the maintenance option menu. The List option always begins with the current record. If it is necessary to List all records, use the Beg option to make the first record the current record, and then choose List.

FIGURE 3.8

Summary of cost control system data entry rules.

When you are sure that the data entry is correct, return to the cost control system main menu. Check the results of your work by doing an INQUIRY BY MONTH for March and an ENGINEERS' ACTIVITY ANALYSIS for Sarah Ludwig. Print both G/L Budget vs. Actual Reports and both Job Cost Reports using a cutoff date of 03/31/89. Also print both Expense History Reports. The results are shown in Figures 3.9 through 3.14.

```
DATE: XX/XX/XX                                                   PAGE:    1
TIME:   XX:XX
                        EXPENSES BY MONTH FOR March, 1989

   SOURCE                ACCOUNT                JOB  DESCRIPTION           AMOUNT HOURS

   ADAMS SUPPLY          4200 MATERIALS         B107 PROTOTYPE MATERIALS   4000.00     0
   ETW LEASING           4300 EQUIPMENT         B107 CAD COMPUTER SYSTEM   4500.00     0
   ASPEN ENGINEERING     4400 SUBCONTRACTORS    B107 CLEAN AIR DESIGN      5000.00     0
   NAL-TECH              4200 MATERIALS         A141 TEST MATERIALS        1500.00     0
   ADAMS SUPPLY          4200 MATERIALS         B762 FOUNDATION BLOCKS     2000.00     0
   BOB JONES             4100 ENGINEERING       A141 DESIGN                4950.00    55
   SARAH LUDWIG          4100 ENGINEERING       A141 STRUCTURAL TESTING    1200.00    30
   SAM TILDEN            4100 ENGINEERING       B107 CONSULTATION          2400.00    40
   BOB JONES             4100 ENGINEERING       B762 DESIGN                 990.00    11
   SARAH LUDWIG          4100 ENGINEERING       B762 DESIGN                2400.00    60

   ** Grand total                                                        28940.00   196
```

FIGURE 3.9

**Expense INQUIRY BY MONTH for month
ending 03/31/89.**

```
DATE: XX/XX/XX                                                   PAGE:    1
TIME:   XX:XX
                        ENGINEERS' ACTIVITY ANALYSIS
                        ANALYSIS BY ENGINEER FOR LUDWIG

   DATE        SOURCE            DESCRIPTION            JOB    HOURS        %

   01/25/89    SARAH LUDWIG      DESIGN                 B762     15         5
   02/25/89    SARAH LUDWIG      DESIGN                 B762     40        14
   03/25/89    SARAH LUDWIG      DESIGN                 B762     60        22

     * Subtotal for DESIGN                                      115        42

   01/25/89    SARAH LUDWIG      STRUCTURAL TESTING     A141     75        27
   02/25/89    SARAH LUDWIG      STRUCTURAL TESTING     A141     50        18
   03/25/89    SARAH LUDWIG      STRUCTURAL TESTING     A141     30        11

     * Subtotal for STRUCTURAL TESTING                          155        57

     ** Grand total                                             270       100
```

FIGURE 3.10

**ENGINEERS' ACTIVITY ANALYSIS through
March 1989 for Sarah Ludwig.**

```
DATE: XX/XX/XX                                                PAGE:    1
TIME:   XX:XX
                          HORATIO & CO.
                       COST CONTROL SYSTEM
                G/L BUDGET VS ACTUAL REPORT MONTHLY

                       THROUGH: 03/31/89


ACCOUNT                      BUDGET        ACTUAL      VARIANCE

4100 ENGINEERING            12000.00      11940.00       60.00
4200 MATERIALS               7500.00       7500.00        0.00
4300 EQUIPMENT               5000.00       4500.00      500.00
4400 SUBCONTRACTORS          5000.00       5000.00        0.00

Grand total                 29500.00      28940.00      560.00




DATE: XX/XX/XX                                                PAGE:    1
TIME:   XX:XX
                          HORATIO & CO.
                       COST CONTROL SYSTEM
              G/L BUDGET VS ACTUAL REPORT YEAR TO DATE

                       THROUGH: 03/31/89


ACCOUNT                      BUDGET        ACTUAL      VARIANCE

4100 ENGINEERING            36000.00      37440.00    -1440.00
4200 MATERIALS              22500.00      15000.00     7500.00
4300 EQUIPMENT              15000.00      13500.00     1500.00
4400 SUBCONTRACTORS         15000.00      17500.00    -2500.00

Grand total                 88500.00      83440.00     5060.00
```

FIGURE 3.11

**G/L Budget vs. Actual Reports through
03/31/89.**

```
DATE: XX/XX/XX                                                   PAGE:    1
TIME:   XX:XX
                            HORATIO & CO.
                         COST CONTROL SYSTEM
                       JOB COST REPORT MONTHLY

                THROUGH: 03/31/89   DEPARTMENT: Engineering


                        JOB                  AMOUNT

                        A141                 7650.00
                        B107                15900.00
                        B762                 5390.00


                        Grand total         28940.00
```

```
DATE: XX/XX/XX                                                   PAGE:    1
TIME:   XX:XX
                            HORATIO & CO.
                         COST CONTROL SYSTEM
                    JOB COST REPORT PROJECT TO DATE

                THROUGH: 03/31/89   DEPARTMENT: Engineering


                        JOB                  AMOUNT

                        A141                26750.00
                        B107                43700.00
                        B762                12990.00


                        Grand total         83440.00
```

FIGURE 3.12

Job Cost Reports through 03/31/89.

```
DATE: XX/XX/XX                                            PAGE:    1
TIME:   XX:XX
                           HORATIO & CO.
                        COST CONTROL SYSTEM
                  EXPENSE HISTORY REPORT BY ACCOUNT


DATE        SOURCE          DESCRIPTION          JOB     AMOUNT

* ACCOUNT: 4100-ENGINEERING

** January
01/25/89 BOB JONES          DESIGN               A141    3600.00
01/25/89 BOB JONES          DESIGN               B762    3600.00
01/25/89 SARAH LUDWIG       STRUCTURAL TESTING   A141    3000.00
01/25/89 SARAH LUDWIG       DESIGN               B762     600.00
01/25/89 SAM TILDEN         CONSULTATION         B107    2400.00
** Subtotal for month                                   13200.00

** February
02/25/89 BOB JONES          DESIGN               A141    4500.00
02/25/89 SARAH LUDWIG       STRUCTURAL TESTING   A141    2000.00
02/25/89 SAM TILDEN         CONSULTATION         B107    2400.00
02/25/89 BOB JONES          DESIGN               B762    1800.00
02/25/89 SARAH LUDWIG       DESIGN               B762    1600.00
** Subtotal for month                                   12300.00

** March
03/25/89 BOB JONES          DESIGN               A141    4950.00
03/25/89 BOB JONES          DESIGN               B762     990.00
03/25/89 SAM TILDEN         CONSULTATION         B107    2400.00
03/25/89 SARAH LUDWIG       STRUCTURAL TESTING   A141    1200.00
03/25/89 SARAH LUDWIG       DESIGN               B762    2400.00
** Subtotal for month                                   11940.00


* Subtotal for account                                  37440.00

* ACCOUNT: 4200-MATERIALS

** January
01/13/89 NAL-TECH           TEST MATERIALS       A141    3500.00
** Subtotal for month                                    3500.00
```

FIGURE 3.13

**Expense History Report by Account through
03/31/89 (page 1 of 3).**

```
** February
02/01/89 ADAMS SUPPLY      PROTOTYPE MATERIALS   B107   1500.00
02/10/89 NAL-TECH          TEST MATERIALS        A141   2500.00
** Subtotal for month                                   4000.00

** March
03/01/89 ADAMS SUPPLY      PROTOTYPE MATERIALS   B107   4000.00
03/10/89 NAL-TECH          TEST MATERIALS        A141   1500.00
03/15/89 ADAMS SUPPLY      FOUNDATION BLOCKS     B762   2000.00
** Subtotal for month                                   7500.00

* Subtotal for account                                 15000.00

* ACCOUNT: 4300-EQUIPMENT

** January
01/01/89 ETW LEASING       CAD COMPUTER SYSTEM   B107   4500.00
** Subtotal for month                                   4500.00

** February
02/01/89 ETW LEASING       CAD COMPUTER SYSTEM   B107   4500.00
** Subtotal for month                                   4500.00

** March
03/01/89 ETW LEASING       CAD COMPUTER SYSTEM   B107   4500.00
** Subtotal for month                                   4500.00

* Subtotal for account                                 13500.00

* ACCOUNT: 4400-SUBCONTRACTORS

** January
01/01/89 ASPEN ENGINEERING CLEAN AIR DESIGN      B107   5000.00
** Subtotal for month                                   5000.00

** February
02/01/89 ASPEN ENGINEERING CLEAN AIR DESIGN      B107   5000.00
02/01/89 ASPEN ENGINEERING CAD PROGRAMMING       B107   2500.00
** Subtotal for month                                   7500.00
```

FIGURE 3.13

**Expense History Report by Account through
03/31/89 (page 2 of 3).**

```
** March
03/01/89 ASPEN ENGINEERING CLEAN AIR DESIGN     B107    5000.00
** Subtotal for month                                   5000.00

* Subtotal for account                                 17500.00


*** Grand total                                        83440.00
```

FIGURE 3.13

**Expense History Report by Account through
03/31/89 (page 3 of 3).**

```
DATE: XX/XX/XX                                            PAGE:     1
TIME:   XX:XX
                            HORATIO & CO.
                         COST CONTROL SYSTEM
                     EXPENSE HISTORY REPORT BY JOB

    DATE      SOURCE           ACCOUNT          DESCRIPTION          AMOUNT

    ** JOB A141

    * January
    01/13/89 NAL-TECH          4200-MATERIALS   TEST MATERIALS        3500.00
    01/25/89 BOB JONES         4100-ENGINEERING DESIGN                3600.00
    01/25/89 SARAH LUDWIG      4100-ENGINEERING STRUCTURAL TESTING     3000.00
    * Subtotal for month                                             10100.00

    * February
    02/10/89 NAL-TECH          4200-MATERIALS   TEST MATERIALS        2500.00
    02/25/89 BOB JONES         4100-ENGINEERING DESIGN                4500.00
    02/25/89 SARAH LUDWIG      4100-ENGINEERING STRUCTURAL TESTING     2000.00
    * Subtotal for month                                              9000.00

    * March
    03/10/89 NAL-TECH          4200-MATERIALS   TEST MATERIALS        1500.00
    03/25/89 BOB JONES         4100-ENGINEERING DESIGN                4950.00
    03/25/89 SARAH LUDWIG      4100-ENGINEERING STRUCTURAL TESTING     1200.00
    * Subtotal for month                                              7650.00
```

FIGURE 3.14

**Expense History Report by Job through
03/31/89 (page 1 of 2).**

```
** Subtotal for job                                                26750.00

** JOB B107

* January
01/01/89 ETW LEASING        4300-EQUIPMENT      CAD COMPUTER SYSTEM  4500.00
01/01/89 ASPEN ENGINEERING  4400-SUBCONTRACTORS CLEAN AIR DESIGN     5000.00
01/25/89 SAM TILDEN         4100-ENGINEERING    CONSULTATION         2400.00
* Subtotal for month                                                11900.00

* February
02/01/89 ETW LEASING        4300-EQUIPMENT      CAD COMPUTER SYSTEM  4500.00
02/01/89 ASPEN ENGINEERING  4400-SUBCONTRACTORS CLEAN AIR DESIGN     5000.00
02/01/89 ASPEN ENGINEERING  4400-SUBCONTRACTORS CAD PROGRAMMING      2500.00
02/01/89 ADAMS SUPPLY       4200-MATERIALS      PROTOTYPE MATERIALS  1500.00
02/25/89 SAM TILDEN         4100-ENGINEERING    CONSULTATION         2400.00
* Subtotal for month                                                15900.00

* March
03/01/89 ETW LEASING        4300-EQUIPMENT      CAD COMPUTER SYSTEM  4500.00
03/01/89 ASPEN ENGINEERING  4400-SUBCONTRACTORS CLEAN AIR DESIGN     5000.00
03/01/89 ADAMS SUPPLY       4200-MATERIALS      PROTOTYPE MATERIALS  4000.00
03/25/89 SAM TILDEN         4100-ENGINEERING    CONSULTATION         2400.00
* Subtotal for month                                                15900.00

** Subtotal for job                                                 43700.00

** JOB B762        .

 * January
01/25/89 BOB JONES          4100-ENGINEERING    DESIGN               3600.00
01/25/89 SARAH LUDWIG       4100-ENGINEERING    DESIGN                600.00
* Subtotal for month                                                 4200.00

* February
02/25/89 BOB JONES          4100-ENGINEERING    DESIGN               1800.00
02/25/89 SARAH LUDWIG       4100-ENGINEERING    DESIGN               1600.00
* Subtotal for month                                                 3400.00

* March
03/15/89 ADAMS SUPPLY       4200-MATERIALS      FOUNDATION BLOCKS    2000.00
03/25/89 BOB JONES          4100-ENGINEERING    DESIGN                990.00
03/25/89 SARAH LUDWIG       4100-ENGINEERING    DESIGN               2400.00
* Subtotal for month                                                 5390.00

** Subtotal for job                                                 12990.00

*** Grand total                                                     83440.00
```

FIGURE 3.14

**Expense History Report by Job through
03/31/89 (page 2 of 2).**

Changing and Deleting Expenses

The Del and Modify options are used to delete and change records, respectively. Both options work on the current record only, so it is usually necessary to use the searching options such as List, Ret, Next, and Prev to make the record to be deleted or modified the current record.

A simple way to do this is to List the records from the first record to identify the one to be deleted or modified. Use the Ret option and enter the date of the desired record. Press PgDn to activate the retrieve function. If the desired record is displayed, choose Del or Modify accordingly. If the desired record is not the current record, use Next and/or Prev to find the desired record, and then choose Del or Modify accordingly.

For practice change the March 10 transaction for Account 4200 and Job A141 from $1500 to NAL-TECH for TEST MATERIALS to $2800 to NAL-TECH for TEST MATERIALS. Also delete the February 1 transaction for Account 4200 and Job B107 for $1500 to ADAMS SUPPLY for PROTOTYPE MATERIALS.

When you have completed these tasks, choose the Quit option to return to the cost control system main menu. The system will ask you if you want to reorganize the database. Press Y to answer yes. As part of the update process, the cost control system allows the user to reorganize the database whenever a record has been deleted. Database reorganization is discussed in Chapter 9.

To verify the results of your work, do an INQUIRY BY ACCOUNT for Account Number 4200, and compare the display to the Expense History Report by Account in Figure 3.13. The inquiry is shown in Figure 3.15.

Notice that you now have only one February transaction and that the second transaction for March reads $2800. The effect on the total for Account 4200 is a net decrease of $200 from $15000 in the previous Expense History Report by Account, Figure 3.13, to $14800 in the inquiry. This occurred because you deleted a $1500 transaction and increased an existing transaction by $1300, causing a $200 net decrease in the total.

Purging Records

The cost control system is designed to maintain expense and budget records for two years. At the end of each month the system operator runs option 6, PURGE JOBS, from the cost control system main menu, Figure 2.1. This module removes all expense and budget records for jobs that have not had any activity for two years. The PURGE JOBS option is included on the cost control system main menu, Figure 2.1, but the program has been deactivated from the version of the system you received. Try it.

```
DATE: XX/XX/XX                                              PAGE:    1
TIME:   XX:XX
                        EXPENSES BY ACCOUNT FOR 4200

   SOURCE              ACCOUNT          JOB  DESCRIPTION        AMOUNT   HOURS


* January
NAL-TECH               4200 MATERIALS   A141 TEST MATERIALS    3500.00      0
* Subtotal for month
                                                              3500.00      0

* February
NAL-TECH               4200 MATERIALS   A141 TEST MATERIALS    2500.00      0
* Subtotal for month
                                                              2500.00      0

* March
ADAMS SUPPLY           4200 MATERIALS   B107 PROTOTYPE MATERIALS 4000.00    0
NAL-TECH               4200 MATERIALS   A141 TEST MATERIALS    2800.00      0
ADAMS SUPPLY           4200 MATERIALS   B762 FOUNDATION BLOCKS 2000.00      0
* Subtotal for month                                          8800.00      0

** Grand total                                               14800.00      0
```

FIGURE 3.15

**Inquiry on Account 4200 after Change/
Delete Expenses.**

Since data stores cannot grow indefinitely, every system must somehow provide a purging operation. The choice of purging method depends upon the type of system and how the system is used.

The cost control system is a transaction processing system that maintains a BUDGETS master data store and an EXPENSES transaction data store in order to produce reports and inquiries. The cost control system provides detailed reports such as the Expense History Reports and summary reports such as the Job Cost Reports.

In systems that do not provide detailed reports there is less need to maintain transaction records. Transactions may be summarized during the update function, and the summarized information recorded in the master data store. Once the transactions are summarized, they might be deleted

automatically or held for a short time before deletion. This is another way
to provide the purging operation.

Master Maintenance

A realistic transaction processing system must give the users the ability to
add, change, and delete master data store records. For instance, cost control
system users must enter the budget figures for a new month to produce
accurate Budget vs. Actual Reports.

To perform master file maintenance, choose option 1, MAINTAIN
BUDGETS, from the cost control system main menu, Figure 2.1. Since the
BUDGETS data store is made up of two files, a second menu is presented.
Option 1, MAINTAIN ACCOUNTS, is used to maintain the account identi-
fication file, and option 2, MAINTAIN MONTHLY BUDGET FIGURES, is
used to maintain the budget file. Choose option 2.

This option works the same way as MAINTAIN EXPENSES in the cost
control system. For practice use List to determine the budget figures for
each account/job combination in March, and then use Add to enter the
identical figures for each account/job combination in April.

When this is done, enter some April EXPENSES and trace their effect
through INQUIRIES and REPORTS. Do not be afraid to experiment. If you
are using the educational version of dBASE III PLUS, you will not be able to
enter more than three April expense transactions without deleting prior
transactions. If you are using the commercial version of dBASE III PLUS,
then you may enter as many transactions as you like.

Summary of Transaction Processing and/or Reporting Systems

The operations of a transaction processing and/or reporting system are
summarized in Figure 3.16. The figure represents an expanded version of
Figure 3.1.

Most accounting applications are transaction processing and/or reporting systems that fit this model. The details vary, but the general design is
the same. Figure 3.17 lists the transaction and master files for several common accounting applications.

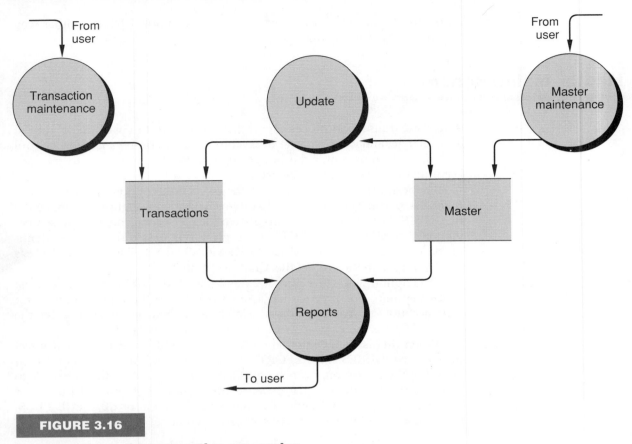

FIGURE 3.16

The operations of a transaction processing and/or reporting system.

Summary

This chapter presents a look at the transaction processing component of the cost control system. Through this component, users maintain the system's database of expenses and budgets. This database supports the inquiries and reports that were presented in Chapter 2.

Chapter 3 completes Part One of this text. The chapters of Part Two look back in time to the days before the cost control system was developed. They present the conditions that led to the development of the system, and they trace the path followed by the development team through the early phases of the life cycle.

APPLICATION	MASTER	TRANSACTIONS
Inventory control	Items Chart of accounts	Sales Deliveries
Accounts receivable	Customers Items Chart of accounts	Sales Cash receipts
Accounts payable	Vendors Items Chart of accounts	Purchases Cash disbursements
General ledger	Chart of accounts	Journal entries Other applications' transactions

FIGURE 3.17

Common accounting systems.

THOUGHT QUESTIONS
In Your Opinion. . .

1. In MAINTAIN EXPENSES how does the data entry process for Account Number differ from the data entry process for Job Number? What are the ramifications of that difference? Should one of the data entry processes be changed?

2. In MAINTAIN EXPENSES how does the data entry process for Account Number differ from the data entry process for Source and Description? What are the ramifications of that difference? Should one of the data entry processes be changed?

3. Why do you think the decision was made to retain expense transactions for two years after the close of a job? Are any reports adversely affected by this decision? Are any inquiries adversely affected by this decision? What can be done to remove the adverse effects of this decision on the reports and inquiries?

HORATIO&CO.

Day-to-Day Activity

The credit union system is also a transaction processing system. The master data store is called MASTER. It is made up of one file that contains one record for each account holder; each record contains the number of the account, the name of the account holder, his or her address and city, state, and zip code, the opening account balance for the current month, the current balance, the interest paid year to date, and a maintenance fee indicator that tells if the account balance has fallen below $1000 this month. Figure 3.1A shows the current contents of MASTER.

The transaction data store is a single file called TRANS. Each record represents a deposit to or a withdrawal from an account. Each record contains the account number, the code for the teller recording the transaction, the date, an indicator showing whether the transaction is an original or a reversal of an existing transaction (O/R), the type of transaction—deposit or withdrawal (D/W), the amount, and an indicator showing whether cash or check was used for the transaction (C/K). Figure 3.2A shows the current contents of TRANS.

The following sections describe the transaction maintenance, purging, and master maintenance modules of the credit union system. Notice the similarities to and differences from the cost control system modules.

NUMBER	NAME	ADDRESS	CITY, ST, ZIP	OPENING BALANCE	CURRENT BALANCE	INTEREST	MAINT FEE INDICATOR
K101	MARY KILPATRICK	4 EAST 88 STREET	NEW YORK, NY 10010	329.67	1453.24	34.17	.T.
D423	SAM DUVALL	10 TENALY ROAD	FORT LEE, NJ 07655	4526.77	4526.77	26.77	.F.
S097	FRANCIS SILVA	1148 39 STREET	BROOKLYN, NY 11218	27498.40	26513.40	248.40	.F.
B313	ERIKA BERG	58 VICTORY PLACE	NUTLEY, NJ 07213	1855.91	2930.91	13.41	.F.

FIGURE 3.1A

Listing of MASTER data.

NUMBER	TELLER	DATE	ORIG/ REVERSAL	TYPE	AMOUNT	CASH/ CHECK
K101	2Q3	07/10/88	O	D	100.00	K
K101	774	07/22/88	O	D	100.00	C
K101	774	07/27/88	O	W	575.00	C
D423	2Q3	07/15/88	O	D	200.00	C
D423	774	07/22/88	O	D	100.00	K
B313	774	07/05/88	O	D	500.00	C
B313	774	07/10/88	O	W	1000.00	C
B313	774	07/22/88	O	W	400.00	C
B313	774	07/28/88	O	W	400.00	C
B313	774	08/05/88	O	W	547.50	C
K101	774	08/05/88	O	W	5000.00	K
S097	774	08/05/88	O	D	4250.00	K
B313	2Q3	08/20/88	O	D	1550.00	C
D423	774	08/05/88	O	D	3000.00	K
S097	774	09/13/88	O	W	985.00	C
B313	774	09/13/88	O	D	675.00	C
K101	774	09/13/88	O	D	345.69	K
B313	774	09/23/88	O	D	400.00	K
K101	2Q3	09/23/88	O	D	777.88	C
B313	2Q3	09/30/88	O	D	100.00	C
D423	2Q3	09/30/88	O	W	100.00	C
B313	2Q3	09/30/88	R	D	100.00	C
D423	2Q3	09/30/88	R	W	100.00	C

FIGURE 3.2A

Listing of TRANS data.

Maintaining Transactions

The transaction maintenance option of the credit union system is option 2, MAINTAIN TRANSACTIONS (see Figure 2.1A). Walter uses this option to record deposits and withdrawals. When Walter is on lunch break or otherwise unavailable, Camille fills in for him at the teller's window. The transaction maintenance menu is shown in Figure 3.3A. Notice that transactions cannot be changed or deleted. Changes are accomplished by entering a reversal for the erroneous transaction and reentering the correct data. Deletions are accomplished through a reversal only. (Why do you think the system was designed this way?)

```
MAINTAIN TRANSACTIONS
---------------------


                    1. ADD TRANSACTIONS

                    2. LIST TRANSACTIONS

                    3. END
```

MAINTAIN TRANSACTIONS menu.

Something New

Option 5, CLOSE OUT MONTH, of the credit union system main menu, Figure 2.1A, is run at the end of the month. The cost control system did not contain a similar module. This program records the monthly interest and maintenance fees shown on the MONTHLY STATEMENTS in the account holder's MASTER record. No matter how many times the monthly reports are printed, the interest and maintenance amounts are not recorded in the account records until this program is run. This allows for the possibility of difficulties with the printer, such as paper jams, in the MONTHLY STATEMENT run. If the account records were being updated during the print, then restarting the print program would require setting the updated records back to their original state.

In addition to recording interest and maintenance fees, the CLOSE OUT MONTH module sets the account record opening balance for the month equal to the current balance plus interest minus maintenance fee if any. Finally, the new month is recorded as the system's current month for programs such as ACCOUNT INQUIRY.

Purging Records

Option 6, END OF YEAR, of the credit union system main menu, Figure 2.1A, provides the purging function. After interest earned is reported to the IRS by the YEAR END 1099S program, it is erased from the MASTER records by this program to begin the new year. All transactions for the previous year are copied to an archive and erased so that the system begins the new year with a clean slate. This program has been deliberately deactivated in the version of the credit union system you received.

Master Maintenance

Option 1, MAINTAIN ACCOUNTS, of the credit union system main menu,
Figure 2.1A, is the master file maintenance module of the credit union
system. It is run whenever a new account holder is added to MASTER or
whenever MASTER information about a current account holder needs to be
changed.

The option is password protected. (Why?) The password is ROSEBUD.
As usual, password characters are not displayed on the screen when they
are typed at the keyboard.

Exercises

1. The current month for the credit union system is September 1988. Run
    option 5, CLOSE OUT MONTH, of the credit union system main menu.
    Enter the digits 06 with no decimal point and no Enter key in response
    to the prompt to enter the annual percentage rate. Wait for the main
    menu to return.

    When the main menu returns, do an ACCOUNT INQUIRY on
    number S097. Notice that only a balance forward line is displayed. This
    is because you have begun a new month, October, and you have not yet
    entered any transactions.

    Also notice that the balance forward amount matches the ending
    balance after interest and maintenance fee for this account from the
    September MONTHLY STATEMENT. Another way to verify that the
    CLOSE OUT MONTH worked properly is to run the TOTAL ON-DE-
    POSIT REPORT and check the balances of each account.

    Repeat exercise 1 for account number K101.

2. Run option 2, MAINTAIN TRANSACTIONS, from the credit union
    system main menu.

    For starters, choose option 2, LIST TRANSACTIONS, from the
    MAINTAIN TRANSACTIONS menu, Figure 3.3A. You will see a list-
    ing of the deposits and withdrawals recorded so far. Notice you have
    transactions for July, August, and September; no transactions have been
    entered for October yet.

    Also notice the information recorded for each transaction: the
    TELLER's code, the account NUMBER, the DATE, the ORIGINAL/
    REVERSAL indicator (O=Original, R=Reversal), the TYPE of transac-
    tion (D=Deposit, W=Withdrawal), the AMOUNT of the transaction,
    and a CASH/CHECK indicator (C=Cash, K=Check).

When you return to the MAINTAIN TRANSACTIONS menu, choose option 1, ADD TRANSACTIONS. Enter the following transaction.

```
        NUMBER: D423
        TELLER: 774
          DATE: 100788—no slashes, no Enter key.
 O/R INDICATOR: O
          TYPE: D
        AMOUNT: 255.78—enter the decimal point, no Enter key.
   CASH/CHECK: C
```

Choose option 2, LIST TRANSACTIONS, to see what you have entered.

To see the effect of your work, return to the credit union system main menu and choose option 3, SCREEN INQUIRIES. Do an ACCOUNT INQUIRY on number D423 and notice the deposit on October 7 for $255.78.

3. Enter the following transactions.

NUMBER	TELLER	DATE	ORIG/ REVERSAL	TYPE	AMOUNT	CASH/ CHECK
B313	774	10/07/88	O	D	489.00	K
S097	774	10/07/88	O	W	5000.00	K
K101	774	10/07/88	O	D	1545.77	C
B313	2Q3	10/20/88	O	W	200.00	C
D423	2Q3	10/20/88	O	W	100.00	C
S097	774	10/20/88	O	D	1200.00	K
K101	774	10/20/88	O	D	234.34	K

List the transactions to check the accuracy of your work and do the proper ACCOUNT INQUIRIES and CASH DRAWERS INQUIRIES to see the results of your work. Print the TOTAL ON-DEPOSIT REPORT, the proper DAILY ACTIVITY LOGS, and the DAILY INTEGRITY REPORT for 102088 to see the results of your work.

4. If a transaction is incorrect, then reversal transactions are used to correct the error. A reversal transaction is one that matches the original in all fields except the ORIGINAL/REVERSAL INDICATOR. O=Original and R=Reversal.

For practice change the withdrawal on October 20 from account B313 for $200 to a check deposit for $200 to the same account on the same day. Enter a reversal for the original first, then enter the new transaction. Do an ACCOUNT INQUIRY to verify the result.

Also delete the deposit on October 7 from account K101 for $1545.77 by entering a reversal for the original transaction. Do an ACCOUNT INQUIRY to verify the result.

5. Close out the month of October. Run MONTHLY STATEMENTS from the PRINTED REPORTS menu, then run CLOSE OUT MONTH from the credit union system main menu. Use 07 as the annual rate for the month. Run TOTAL ON-DEPOSIT REPORT to verify the results.

6. Add a new account for Christine Dine at 404 Madison Avenue, New York, NY 10020. Use account number D424. Christine wants to open her account by depositing a $5000 check.

Systems Analysis

Part Two contains four chapters that cover the analysis phases of the systems development life cycle. Chapter 4 presents a method of analyzing business problems and/or opportunities and determining the potential of an information systems solution. Chapter 5 presents a set of tools and techniques that can be used to refine the prior analysis into a set of requirements for the proposed system. Chapters 6 and 7 present a set of tools and techniques for generating and evaluating solution alternatives that satisfy the system requirements.

The correct determination of requirements is crucial to the delivery of a satisfactory system. In this work the systems analyst calls upon his/her oral and written communication skills and general business knowledge. The technical activities of the later phases of the systems development life cycle are productive only when the analysis phases are properly carried out.

CHAPTER 4

Problem and/or Opportunity Analysis

Objectives

In Chapters 2 and 3, you studied the Horatio & Co. Cost Control System from the user's point of view. Your responsibilities were those of the user: data entry, report generation, and control of data integrity. In this chapter you will study *why* the cost control system was developed. The remainder of the text will explain *how* the system was developed.

Computer-based information systems are designed and built to support the business objectives of the users. Before the systems analyst thinks about technical items such as hardware and software, he or she must communicate with the users in business terms. The analyst must understand the problems and/or opportunities facing the user/management group before solutions and implementations of those solutions are discussed.

In this chapter you will learn how the engineering department controlled costs before the implementation of the automated system you studied in Chapters 2 and 3. You will also learn a method to analyze user problems and/or opportunities. You will be on your way to becoming a systems analyst because accurate problem and/or opportunity analysis is the first step of the systems development life cycle, Figure 1.3.

The specific objectives of the chapter are

1. To determine the procedures and decisions of the Problem and/or Opportunity Analysis Phase of the systems development life cycle;

2. To examine the situation faced by Sam Tilden before the implementation of the cost control system;

3. To trace the activities leading to Sam's decision to initiate the development of the cost control system.

Introduction

Sam Tilden came to Horatio & Co. after earning a BS degree in civil engineering. After three years the company president, Frank Chapin, encouraged him to enroll in an evening MBA program at a nearby university. Mr. Chapin explained that the construction business was becoming more competitive and that better-managed firms were beating Horatio & Co. in the marketplace despite the superior technical expertise of the Horatio staff.

As a result of the increased competition, Horatio & Co. could expect lower revenues in the future. To maintain the current level of profit, costs would have to be cut. To implement this plan, Horatio & Co. was investing in the development of the business skills of its department managers. The president informed Sam that he would be promoted to department manager upon completion of his MBA; the current department manager would return to full-time engineering duties.

Sam earned his MBA and became engineering department manager ten months ago. He began the job with the improvement of the business operations of the department as his top priority.

The Department Sam Inherited

The previous department manager was a fine engineer, but he was not particularly capable as a manager. He considered planning, organizing, accounting, and budgeting activities to be a chore, and he spent a minimum amount of time on them. He was happy to step down and let Sam take over.

The accounting department of Horatio & Co. used a minicomputer to process general ledger, accounts payable, accounts receivable, and payroll activities. Each year the engineering department manager prepared an annual budget based upon signed construction contracts. By the 20th of each month the accounting department distributed an updated annual Budget vs. Actual Report showing year-to-date actual expenses through the end of the previous month, annual budget, and remaining balance. The report was printed in general ledger account number sequence.

The previous engineering department manager was happy with this report. He kept a copy of every payment authorization entered into the accounts payable or payroll system. Every month he used these copies to check the Budget vs. Actual Report for accuracy and filed the report away.

The Budget vs. Actual Report provided no help with tracking costs by job number; a manual ledger was used for this purpose. Every time a payment authorization was entered into the accounts payable or payroll systems, the previous manager would record the date, source, description, and amount of the expenditure on the page of the ledger corresponding to the job for which the expense was incurred. At the end of the month these figures were totaled and passed to the project managers for billing to the clients. Since budgets by job were not prepared, no comparison of actual costs to budget figures at the job level was possible.

Sam was unhappy with the job cost system. He felt it forced him to use staff-time inefficiently. Once information was entered into the job cost ledger, it was not accessible for other purposes, particularly budget vs. actual analysis. Preparing special reports was a nightmare because the reports required copying data from the ledger, rearranging this data, and computing totals manually. All of the engineers complained about this component of their work.

The annual budget figures by general ledger account did nothing to control costs. Overruns in the early part of the year were not identified until late in the year when year-to-date actual expenses for the account finally exceeded the annual budget figures. In each of the three years Sam had been at Horatio & Co., actual expenses exceeded budget figures in all accounts.

Sam's Plan of Action

Sam identified several ways to improve the business operations of the engineering department. Better access to job cost information could decrease clerical costs and improve the work life of the engineers. A monthly budgeting system by both job and general ledger account could control costs better and identify overruns faster.

Better deals could be negotiated with vendors through a reorganization of the vendor base. By reducing the number of vendors, Horatio's volume with each of the remaining vendors would increase, thereby improving Horatio's status with the vendor.

Turnover could be reduced by instituting professional development plans for the engineers. Employee turnover is a serious problem for a specialized firm such as Horatio & Co. Competent engineers seek varied work assignments to keep their technical skills up to date.

Sam knew he did not have the means or the time to accomplish these tasks. Proper use of a monthly budgeting system would significantly increase the time he spent "crunching numbers." Reorganizing the vendor base would require access to a complete history of Horatio's activity with each vendor. Professional development planning would require regular meetings with the engineers to discuss work assignments and career goals. At these meetings Sam would need a summarized report of the engineer's recent activity so that progress on past plans could be checked and new plans could be made.

As part of his MBA program, Sam took a course in computer-based information systems. His instincts told him that an automated system would improve his efficiency and support the plan he hoped to implement.

Sam contacted his information systems professor, Pete Willard, who agreed to serve as systems analysis consultant. At their first meeting Pete talked about his views on the use of computers in business and the systems development life cycle.

"The use of computers varies from one organization to another," explained Pete. "The extent of computer use depends upon the work of the organization and, more importantly, upon the people who do the work. It is possible for two companies to perform exactly the same work in entirely different ways, so the systems development process must involve learning about both the work and the people of the organization.

"The process begins with the business environment," Pete continued. "Once a problem and/or opportunity has been identified, management must express the objectives of a solution effort. Given objectives, the management group should survey the organization's strengths and weaknesses to develop a set of tactics to achieve the objectives [2].

"Computer-based information systems support business objectives and tactics. A systems analyst's first task is to review management's objectives and tactics and to identify information system functions that support them. If the people involved feel that these functions will support the way they plan to implement the tactics, then they should study an information system solution further. The people involved include both the users and their managers.

"To study an information system solution further, the user/management group and the systems analyst develop systems analysis models of the proposed system and, perhaps, the current system. The extent of the model building depends upon the system and its development and operating environment.

"From the models the analyst outlines several paths to a final operational system for the user/management group. The different paths and final systems involve varying cost, complexity, and development methodology [1].

"If management decides to proceed with an information system solution, the development team undertakes a combination of design, implementation, and evaluation activities. Like model building, the actual activities and the sequence in which they are performed depend upon the development team, the system, and the system's development and operating environment. The design-implement-evaluate activities must consider all five components of a computer-based information system: hardware, software, data, procedures, and personnel" [3].

Pete summarized his ideas in Figure 4.1. "The picture is deceiving," he explained. "The process is not linear. One phase does not follow directly from another. The development team can jump ahead to gather information or loop back to a previous phase based upon information uncovered in a subsequent phase. Many times the team will work on more than one phase simultaneously. The diagram simply presents the activities. Sequence and repetition depend upon the project and the choices made throughout the project."

The Presentation to Management

Together Sam and Pete organized Sam's plan into a statement of business objectives and tactics. An information system to support Sam's plan would have to provide summary and history reports of engineering department expenses. Selective inquiries would also be required. The system would be used for both budgeting and job cost purposes.

Sam was encouraged by the apparent potential of a computer-based information system to support his tactics. Pete estimated that the Model Building Phase and the Evaluation of Alternatives Phase would take five days to complete. The potential benefits seemed worth the risk of this small amount of time and money, so Sam decided to proceed with the next step of the systems development life cycle. He summarized his plan and submitted it to Mr. Chapin. The plan summary appears in Figure 4.2.

The president liked Sam's plan, but he had some reservations. He was particularly disturbed by the fact that Sam did not provide any estimates of the cost savings.

Sam admitted that, at this time, he did not know how much money would be saved. He based his plan on his observations of the department over a number of years. After the Model Building and Evaluation of System Alternatives Phases of the life cycle, estimates would be possible. No additional commitments would be required until those estimates were available.

Mr. Chapin was not entirely satisfied, but he agreed that the potential savings were worth the risk of five days' time. He approved the start of the Model Building and Evaluation Phases. He agreed to write a memo introducing Pete and announcing the start of the project to all Horatio & Co. personnel.

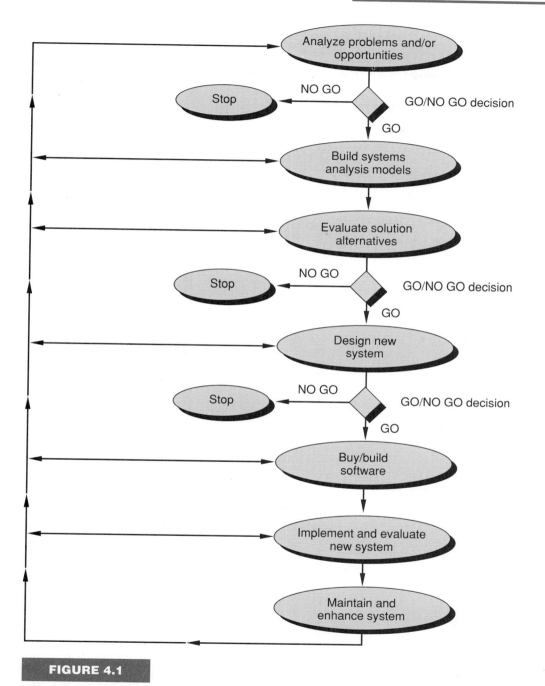

FIGURE 4.1

The systems development life cycle.

Business Objective

Maintain the current level of profit in the engineering department by

      decreasing clerical costs
      identifying cost overruns within 30 days
      decreasing cost of materials
      decreasing turnover

Business Tactics

1. Eliminate manual preparation of job cost reports

2. Institute a monthly budgeting system by job number and general ledger account to replace the current annual system by general ledger account only

3. Reorganize the vendor base for higher volumes with fewer vendors

4. Institute professional development plans for engineering personnel

System Objective

Reports and inquiries on demand including

      Budget vs. Actual Reports by job number and general ledger account on a monthly basis
      Job Cost Summary and Detail Reports
      Detail INQUIRY BY SOURCE of expense or vendor

Next Step

Proceed to the Model Building Phase of the systems development life cycle.

FIGURE 4.2

Sam Tilden's plan to improve business operations in the engineering department.

At Pete's request Mr. Chapin prepared a new organization chart of the company; the previous version was several months out of date. The chart is shown in Figure 4.3.

Pete asked Mr. Chapin to introduce the proposal to Ed Henderson, accounting department manager and the person responsible for the mini-computer-based accounting system, before the release of the memo to the rest of the staff.

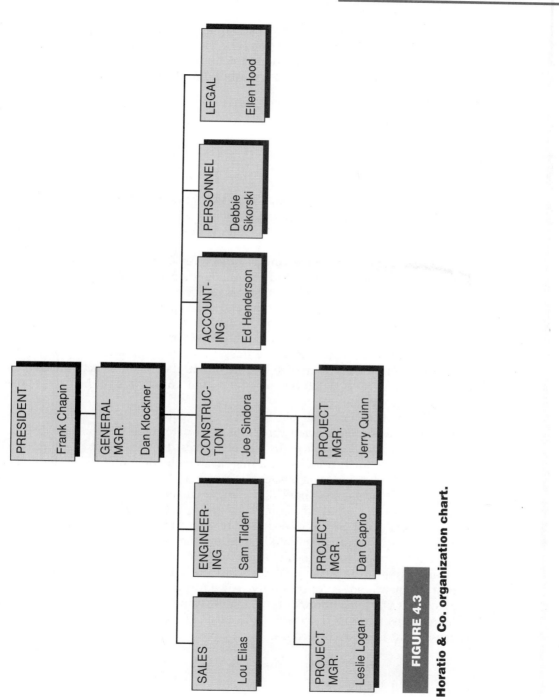

FIGURE 4.3

Horatio & Co. organization chart.

In the remaining sections of this chapter you will look back to study the method Sam and Pete used to analyze the improvement of the business operations of the engineering department. You will see how they came to the decision to proceed to the Model Building Phase of the systems development life cycle.

Business Objectives

Notice the organization of Sam's plan in Figure 4.2. The plan begins with a broadly stated outcome: maintain the current level of profit by decreasing costs. If a new information system is developed, it must support this objective.

The determination of **business objectives** is usually a top-level management activity. You will study methods for determining business objectives in your management and business policy courses.

The systems analyst must know and understand the business objectives of a new information system, but the analyst usually does not participate in the development of the objectives. The development of business objectives requires an understanding of the mission and operation of the company. Since the analyst is not involved in the day-to-day business activities, he or she cannot hope to match the experience and knowledge of the user/management group.

The analyst is responsible, however, for requiring the users to produce an objectives statement that can be used to determine the feasibility of an information system to support the objectives. In reviewing an objectives statement, the analyst looks for a measurable outcome that the system will support. Decreasing costs is such an outcome. Improved return on investment, increased market share, and a larger customer base are others. Objectives should be quantified as early as possible in the process.

At this point in the dialogue with the user/management group, the analyst must keep an open mind regarding the usefulness of an information system in supporting the stated objectives. Business objectives such as decreasing costs or increasing market share are usually developed in response to a problem and/or opportunity. An information system may very well provide a viable solution, but the analyst cannot know this until the user/management group determines the business tactics that will be used to achieve the business objectives.

Business Tactics

Business tactics are actions performed to accomplish business objectives. Sam listed four tactics in his plan, Figure 4.2. Items 1 and 3 involve performance quality and efficiency.

Performance quality refers to how well a job is done. Computer systems often improve performance quality by generating information that is otherwise unavailable. From your work in Chapters 2 and 3 you know that a computer-based information system can help Sam to reorganize his vendor base. Using the INQUIRY BY SOURCE, he will be able to review a complete history of Horatio's activity with a vendor. He will know how long Horatio has been doing business with the vendor, what products were purchased, and how much was spent. Through these inquiries he will be able to identify candidates for negotiation and/or elimination and support his claims with accurate facts. Any price reductions he gains will improve the quality of his performance as department manager.

Producing such information manually requires searching every job cost ledger page for transactions involving a given vendor, copying them to a separate page, arranging the selected transactions in date sequence, and computing a grand total as well as totals by month. If information on another vendor is required, the process must be repeated from the very beginning. This explains why information such as this is not compiled when manual systems are in use.

Performance efficiency measures the output produced from a given set of inputs. Performance efficiency is improved by increasing output, decreasing input, or both. Sam felt he could increase his department's productive output by reducing budgeting and job cost clerical activities.

Computer-based information systems are well suited to improving performance efficiency because they provide better information and require less attention than manual systems.

Users enter transaction data into a computer system only once. After the data is entered, the system provides access for manipulations such as record selection and report generation. If the computer system information replaces manually generated reports, then the staff time saved could be put to more productive use or eliminated. In either case performance efficiency is improved.

To see this more concretely, think about the time it would take to use the EXPENSES file transactions in sequence of original entry, Figure 3.3, to produce the Expense History Report by Job, Figure 2.10, manually.

Business tactics aimed at raising performance quality and efficiency are often used to support business objectives such as decreasing costs. Items 2 and 4 in Sam's plan, however, do not address performance; they deal with the question of **control mechanisms**.

Control mechanisms are used to detect substandard performance. The Budget vs. Actual Reports of the cost control system are examples of control mechanisms. The budget columns represent the expected or standard performance, and the actual columns represent actual performance. The reports compare expected and actual performance and identify undesirable variances.

There are many possible reasons for a variance in the Budget vs. Actual Report. In Chapter 2 you saw that unexpected programming expenses caused a negative variance in Account 4400—SUBCONTRACTORS. Variances in accounts such as MATERIAL or EQUIPMENT might be caused by inaccurate delivery quantities, waste in production, or theft. One goal of an information system control mechanism is the early detection and reporting of the variance.

Sam Tilden will improve the control mechanism of the Budget vs. Actual Report by changing from annual to monthly budgets and introducing budgets by job (Tactic #2). Naturally, more work will be required, so Sam will have to develop procedures and staff for the new tasks. His decision to change was based upon his experience as a Horatio & Co. engineer. He felt that past overruns were detected too late and that budgets only by general ledger account covered too wide a control area to be effective.

The ENGINEERS' ACTIVITY ANALYSIS will help Sam to monitor the engineers' professional development (Tactic #4). Recall Figure 2.13, ENGINEERS' ACTIVITY ANALYSIS for LUDWIG. Sam promised to increase Sarah Ludwig's design activities. The inquiry shows at a glance that progress toward this goal is being made. Sam can feed this information back to Sarah to determine her level of satisfaction, and then make plans for the coming months.

The Decision

Computer-based information systems support business tactics involving performance and control very well. Figure 4.4 presents a summary of the information system functions used for this purpose. These functions represent the tools the systems analyst brings to bear upon user problems and/or opportunities. The user/management group consults this list to decide if an information system represents a viable solution to their problem and/or opportunity. If it does, they proceed to the Model Building Phase of the systems development life cycle.

A CHECKLIST OF INFORMATION SYSTEM FUNCTIONS

Performance Quality and Efficiency

1. Reduce staff time spent on clerical functions leading to more productive use or elimination of that time.
2. Organize and present information quickly.
3. Organize and present information in a variety of ways.
4. Select records from a large database and present them in isolation.
5. Summarize the content of a large number of records into a meaningful report or display.
6. Perform compilations and/or analyses that are otherwise impossible.

Control Mechanisms

1. Monitor processes to detect and report substandard performance.
2. Aid in the determination of the cause of substandard performance.
3. Implement actions to correct substandard performance.

FIGURE 4.4

**Information system functions that support
business tactics involving performance
quality and efficiency, and control
mechanisms.**

When he reviewed Figure 4.4 with Pete, Sam determined that his tactics would make use of all six performance functions and the first two control functions. This analysis led to the decision to proceed to the Model Building Phase of the systems development life cycle.

The ability to develop tactics to support business objectives is the mark of an experienced business person. Managers and systems analysts alike need to develop this skill. While the systems analyst rarely participates in the development of business objectives, he or she should be involved in the development of business tactics. The systems analyst is responsible for educating the user/management group in information system functions, Figure 4.4, and the tactics they support.

The decision on whether or not to support the chosen business tactics with an information system completes the Problem and/or Opportunity Analysis Phase of the systems development life cycle. If a decision in favor of an information system is made, the user/management group, in conjunction with the systems analyst, should formulate a statement of **system objectives** before proceeding to the Model Building Phase of the life cycle. System objectives broadly state the system actions that will support the performance of the required business tactics. See Figure 4.2 for the statement of system objectives for the Cost Control System.

Summary

At this point in your professional development you have one experience with developing business tactics and system objectives. Some of this experience will carry over to other projects, but, for the most part, each problem and/or opportunity is unique. The good systems analyst is one who can synthesize experience into understanding and apply general concepts to the problem at hand.

In the assignment section you will return to the Horatio & Co. Credit Union System. Here you will find a new set of objectives and tactics supported by the same information system functions found in the cost control system. Do the exercises with an eye toward improving your ability to develop business tactics through these functions. Draw upon your experience in Chapters 2 and 3 as users of the cost control and credit union systems.

Use Figure 4.5 as a guide to the Problem and/or Opportunity Analysis Phase, but keep in mind that work at the later steps will often force you to go back to an earlier step to rethink your position.

THE PROBLEM AND/OR OPPORTUNITY PHASE

1.   State the problem and/or opportunity in terms of business objectives.
2.   Decide upon business tactics to support the objectives.
3.   Examine information system functions, Figure 4.4, to support the tactics.
4.   YES/NO decision regarding further study of an information system solution.

   If NO: user/management group seeks alternative solution to problem and/or opportunity.
   If YES: develop statement of system objectives and proceed to the Model Building Phase of the systems development life cycle.

FIGURE 4.5

The procedures and decisions of the Problem and/or Opportunity Analysis Phase of the systems development life cycle.

THOUGHT QUESTIONS
In Your Opinion. . .

1. Basically, the decision to proceed to the next two phases of the life cycle was made because the potential value of the system seemed worth the risk of five days' time. Do you feel this is responsible decision-making? Is it realistic? What alternatives would you suggest?

2. Put yourself in the position of Ed Henderson, accounting department manager. Discuss the ways in which Ed might react to the news of the cost control system development project. How would you react? Discuss how each reaction should be handled by Mr. Chapin, by Sam Tilden, and by Pete Willard.

References

1. Gremillion, L., and P. Pyburn. "Breaking the Systems Development Bottleneck," *Harvard Business Review*, March-April 1983, pp. 130–137.

2. Kozar, K. A. "A User Generated Information System: An Innovative Development Approach," *MIS Quarterly*, Volume 11, Number 2, June 1987, pp. 163–173.

3. Kroenke, D. *Business Computer Systems*. Santa Cruz, CA: Mitchell Publishing Corporation, 1987.

HORATIO&CO.

The Credit Union's Problems and Opportunities

Five years ago, Camille Abelardo started the Horatio & Co. Employee Credit Union on a volunteer basis. The project was such a success that Camille was relieved of half of her responsibilities in the accounting department, and the management of the credit union was included in her formal job description.

The operation of the credit union involved keeping manual records on the 200 members who maintain accounts that are similar to ordinary bank savings accounts. Members made deposits to and withdrawals from their accounts. Credit union windows were open two hours each day; three accounting clerks served as window tellers.

At the end of the month each member received a statement of his or her account showing the opening balance for the month, all deposits and withdrawals, interest earned for the month, maintenance fees charged, and the closing balance for the month.

The center of the manual system was the account card file. The window tellers recorded all deposits and withdrawals on the cards and computed the updated balance.

At the end of the month the three window clerks and two temporary clerks computed the interest earned and entered it on each card. They also charged a maintenance fee of five dollars if the balance of an account fell below $1000 at any time during the month. Finally, a photocopy of the card was mailed to the member as a monthly statement. The credit union window was closed for the three days it took to produce the monthly statements. An example of an account card and the monthly calculations are shown in Figures 4.1A and 4.2A.

In addition to the account cards, the window clerks were required to keep a daily activity log that recorded all of the day's deposits and withdrawals. At the end of each day Camille collected all of the logs to reconcile each clerk's closing cash balance with the opening balance and the day's deposits and withdrawals.

Once the cash balances were verified, deposits from members made by check were prepared for deposit in a bank account held in the name of the credit union. The Union opened each day with $1500 in cash for each win-

CREDIT UNION ACCOUNT CARD

Mary Kilpatrick
4 East 88th Street
New York, New York 10010

DATE	DESCRIPTION	AMOUNT	BALANCE
12/01/86	Balance Forward		5675.50
12/10/86	Deposit	100.00	5775.50
12/22/86	Deposit	100.00	5875.50
12/27/86	Withdrawal	−575.00	5300.50
12/31/86	Interest @ 6% year	28.49	5328.99

FIGURE 4.1A

A sample credit union account card.

Interest Calculation

BALANCE	FROM	TO	# DAYS	BALANCE x # DAYS
5675.50	12/01/86	12/10/86	10	56755.00
5775.50	12/11/86	12/22/86	12	69306.00
5875.50	12/23/86	12/27/86	5	29377.50
5300.50	12/28/86	12/31/86	4	21202.00
				176640.50

AVERAGE DAILY BALANCE = 176640.50 / 31 DAYS = 5698.06

ANNUAL RATE OF INTEREST = .06

INTEREST EARNED = 5698.06 x .06/12 = 28.49

FIGURE 4.2A

**A sample interest calculation using the
average daily balance method.**

dow clerk. If there was excess cash from the day's activity, this was also deposited in the bank account. If cash was needed to start the next day, a withdrawal from the bank account was prepared. Camille made the necessary deposits and/or withdrawals and locked the next day's opening cash in the credit union vault as her final activity of the day.

An example of a daily activity log is shown in Figure 4.3A. Remember that deposits and withdrawals made by check do not affect the cash balance.

At the end of the year the credit union mailed 1099 forms to each member with copies to the Internal Revenue Service. The 1099s showed the member's account number, name and address, and the total interest earned for the year.

Camille's Problem and/or Opportunity

The improvement of the business operations of the engineering department represents a long-term solution to a growing problem. Camille Abelardo, however, faces a different situation. Due to exceptionally rapid growth, her operation must deal with immediate problems in performance, control, and

---

DAILY ACTIVITY LOG

Teller's Initials: WOR 12/13/86

—Deposits—		—Withdrawals—		Cash
Check	Cash	Check	Cash	Balance
				1500.00 Opening
	250.00			1750.00
			1000.00	750.00
1500.00				750.00
155.00				750.00
345.00				750.00
	500.00			1250.00
	300.00			1550.00

FIGURE 4.3A

A sample daily activity log.

customer satisfaction. She is essentially a victim of her own success, a common occurrence in entrepreneurial ventures.

Camille is concerned about the expense of using three accounting clerks for two hours each day to staff the windows. Because of the high volumes, many errors are made in the maintenance of account cards and the preparation of monthly statements. Recently, employees have begun to complain about errors in their statements and long waiting lines at the windows. Since complaints are handled at the teller windows within earshot of people waiting for service, Camille is concerned that employees will eventually lose confidence in the credit union.

Because of the increasing volume, the daily cash reconciliation is taking longer to complete. Camille is finding that her credit union activities are eating into the time she should be spending on other responsibilities.

Exercises

1. Develop a statement of business objectives for Camille's effort to deal with her problem(s). If you think it is necessary, feel free to include more than one objective in your statement.

2. Develop a set of business tactics to support the objectives of exercise 1. Remember not to impose an information system solution on your thinking at this point in the analysis.

3. Use Figure 4.4 to identify information system functions to support the business tactics developed in exercise 2.

4. From your work in Chapters 2 and 3, you know Camille made her decision in favor of an information system. Develop a statement of system objectives for the credit union system you used in Chapters 2 and 3.

5. Develop a solution to support the business tactics of exercise 2 that does not involve a computer-based information system.

6. Develop a solution to support the business tactics of exercise 2 that involves computers but does not involve a transaction processing system.

7. Compare the situations faced by Sam and Camille. Discuss similarities and differences in the problems and/or opportunities as well as the solutions.

CHAPTER 5

The Model Building Phase

Objectives

This chapter introduces you to systems analysis models. The first two models, the Context Data Flow Diagram and the Analysis Data Flow Diagram, are used to describe the current situation and to communicate among development team members.

Once the current system is modeled, Pete and Sam will compare the plan developed in the Problem and/or Opportunity Analysis Phase, Figure 4.2, to the models. This comparison will yield the Requirements Model for the new system and conclude the Model Building Phase.

Good communication is critical to all systems development efforts. It is important in the Horatio & Co. Cost Control System project because the analyst, Pete Willard, knows nothing about the users' business, and the chief user, Sam Tilden, has little experience with computer-based information systems.

With the development of the End-User Computing (EUC) movement, two-person development teams such as the one made up of Sam and Pete have become common. Studying their efforts will provide a good first experience in systems development. As you read the dialogues between Pete and Sam, think about the complications that would arise if more than one analyst or more than one user were involved.

The specific objectives of the chapter are

1. To use the Context Data Flow Diagram and the Analysis Data Flow Diagram to model an existing system;
2. To determine new system requirements from the prior analysis of problems and/or opportunities and the data flow diagram models;
3. To observe the communication that takes place among members of a systems development team.

Determining the Context
of the System

Pete began the meeting with Sam by explaining what he hoped to achieve. "We are considering the development of an information system to support the business objectives and tactics you outlined in your last report to Mr. Chapin (Figure 4.2). The feasibility of the project depends upon what you want the system to do and how well the system can be integrated into the Horatio environment. In these meetings, therefore, we must come to an understanding of the current environment as well as your requirements for the new system. The meetings will continue until we have a shared, complete, and accurate model of the current and desired systems" [2].

Pete drew the diagram shown in Figure 5.1 on a chalkboard. "This is my conception of your cost system," he said. "Customers give you specifications. You assemble the necessary services and materials. The customers pay you, and you pay your suppliers."

Sam struggled to contain a smile. "It's not that simple, Pete," said Sam. "You've got a lot to learn about the construction business."

"That's why I'm here," said Pete. "Show me."

Sam erased the chalkboard. "Actually, we hardly ever see the customers. We deal with the construction project managers at Horatio & Co.; they deal with the customers. When a contract is being negotiated, the project manager sends us the specifications, and I appoint a team of engineers to come up with a set of material and/or service requirements. The requirements include Horatio engineering services as well as the materials and/or services from outside vendors needed to complete the engineering portion of the project.

"I determine the cost of the requirements and send the report back to the project manager. The other departments do the same thing. The

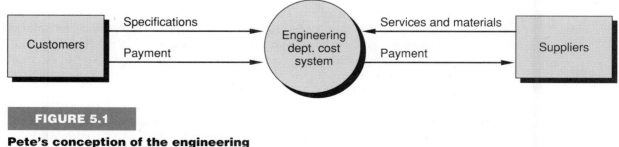

FIGURE 5.1

Pete's conception of the engineering department's cost system.

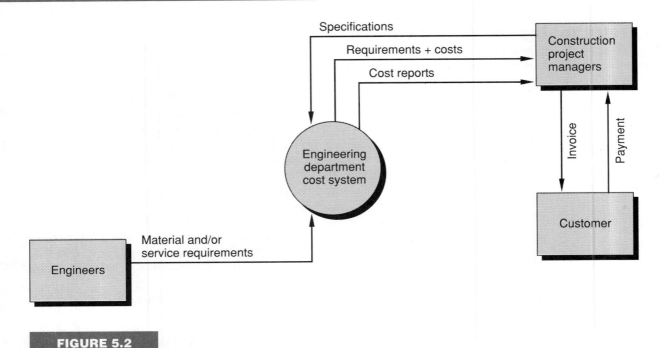

FIGURE 5.2

Sam's first explanation.

project manager uses these estimates to bid the job. Once a project gets underway, we provide regular cost reports to the project managers who bill the clients."

While Sam was talking, he drew the diagram shown in Figure 5.2. "If you do not deal with the customers, forget about them," said Pete. "At this point, we are concerned with items coming into and going out of your office." Sam erased the part of the diagram dealing with customers.

"What goes into the job cost reports?" asked Pete.

"Time and materials," answered Sam. "The engineers use time logs to keep strict account of the time spent on each job. Each week Accounting sends me a preliminary payroll roster showing weekly salary for salaried employees and hourly rate for hourly employees. I use the time logs to fill in the hours worked on the roster. I return the completed roster to Accounting and keep a copy for myself. Betsy Klein, our administrative assistant, enters the completed roster data into the payroll system."

Sam stopped drawing and asked, "How do I show the copy that stays inside the circle and the entry into the computer system?"

"We'll get back to that later," answered Pete. "For now, concentrate on those items that cross the boundaries of your office."

"OK," said Sam. "Betsy uses the rosters and the logs to break down the cost of everybody's time by job and posts the costs in the job cost ledger. At the end of the month I summarize the job cost postings into a Job Cost Report for the project managers, and I use the copies of the rosters to check the Budget vs. Actual Report from Accounting."

Sam's diagram now looked like Figure 5.3. "May I see copies of all of the reports and forms listed on this diagram?" asked Pete.

"Sure," said Sam. "No problem." Figures 5.4 through 5.8 show samples of the reports and forms Pete collected.

"Before we continue, I have a question," said Sam. "You call our operation a system, but a large part of it is manual, and we never had a systems analyst come in and design something for us. Am I missing something?"

"No," answered Pete. "I use the word **system** to describe any process that transforms inputs into outputs. In this project we are attempting to improve your current system of processing cost information. We may replace it with a computer-based system, or we may come up with something else. System is a very general term."

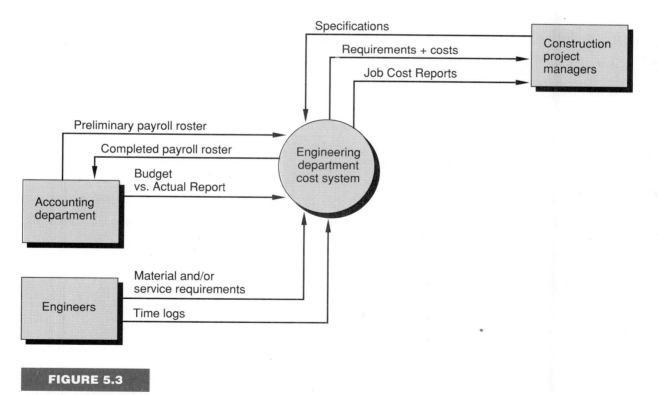

FIGURE 5.3

Sam's expanded explanation.

```
                    MONTHLY JOB COST REPORT
                    ENGINEERING DEPARTMENT
                        APRIL, 1988


        JOB        AMOUNT

        A330      8700.00
        B566      5656.89
        C722      9606.99
        TOTAL    23963.88
```

FIGURE 5.4

A sample Job Cost Report.

Horatio & Co. Weekly Time Log

Name: Bob Jones

Week Ending Date: 4/15/88

Date	Job #: C722		Job # A330		Job #	
	Reg Hrs/ Description	OT Hrs/ Description	Reg Hrs/ Description	OT Hrs/ Description	Reg Hrs/ Description	OT Hrs/ Description
4/11/88	8 Design					
4/12/88	8 Design					
4/13/88	6 Design		2 Testing			
4/14/88	7 Design		1 Testing			
4/15/88	8 Design					

FIGURE 5.5

A sample time log.

Horatio & Co.
Payroll Roster
Week Ending Date 4/15/88

Employee	H/S?	Salary/ Hourly Rate	Hours	Extension
John Chen	S	872.09		872.09
Bob Jones	S	1005.66		1005.66
Sam Tilden	S	930.23		930.23
Tony Long	H	10.00	35	350.00
Dana Carlsen	H	12.50	40	500.00

FIGURE 5.6

A sample payroll roster.

Job Cost Ledger

Job: C722

Date	Source	Account	Amount	Cum. Balance
	Balance Forward			75685.20
4/2/88	Marjam Supply	200 – Materials	3672.00	79357.20
4/10/88	Dial-a-Temp	400 – Subcontractors	200.00	79557.20
4/11/88	Allison Motor	300 – Equipment	867.49	80424.69
4/30/88	Bob Jones	100 – Engineering	4000.00	84424.69
4/30/88	John Chen	100 – Engineering	3750.00	88174.69

FIGURE 5.7

A sample job cost ledger page.

```
                    MONTHLY BUDGET VS ACTUAL REPORT
                            APRIL, 1988

                                       MONTHLY   MONTHLY
        ACCOUNT NAME                    BUDGET    ACTUAL    VARIANCE

            100 ENGINEERING            11500.00 11623.45    -123.45
            200 MATERIALS               7500.00  7500.00       0.00
            300 EQUIPMENT               4000.00  1570.43    2429.57
            400 SUBCONTRACTORS          4300.00  1000.00    3300.00
            500 OVERHEAD                2000.00  2000.00       0.00

                TOTAL                  29300.00 23693.88    5606.12
```

FIGURE 5.8

A sample Budget vs. Actual Report.

"OK," said Sam. "Let me explain how we handle materials. When Accounting receives an invoice from a vendor, they send it to me to authorize payment. I check the purchase order and return a payment authorization voucher. The expense is entered into the accounts payable system and posted in the job cost ledger so that it is included in the Job Cost Report. We keep a copy of the invoice and the voucher to check the Budget vs. Actual Report at month end."

Pete said, "Let's backtrack for a minute, so that I am sure I understand the Job Cost Report. Betsy and you are responsible for processing the engineering department's payroll and accounts payable transactions. Part of that processing involves posting entries into the job cost ledger. At the end of each month the job cost ledger postings are summarized into the Job Cost Report for the project managers."

"You've got it," said Sam.

"Tell me about the Budget vs. Actual Report," said Pete.

"Pretty standard," said Sam. "At the beginning of each year, the department reviews signed construction contracts with the project managers. We use the contracts to determine our expense budget requirements for the year and advise Accounting through a budget authorization sheet signed by the project manager and me.

"As expenses are entered through the accounts payable and payroll systems, our budget lines are charged. Accounting sends us a Budget vs. Actual Report at the end of each month. I use my file of payroll rosters and payment authorization vouchers to check the report. If a contract changes or we get a new contract that affects this year's budget, we notify Accounting through the same countersigned budget authorization sheet."

Pete made an entry on the diagram and asked, "How are purchase orders handled?"

"I need the diagram to explain that," said Sam. He picked up a piece of chalk. "When we receive a new project specification, we send out requests for bids on materials and/or outside services to vendors. I use the best bids to determine the cost of materials and/or outside services; I use payroll rosters to determine the cost of Horatio engineering services. If Horatio gets the contract, the project manager sends us a schedule of the project, and we issue the appropriate purchase orders. When I receive a request for payment authorization, I retrieve my copy of the purchase order and note the payment information on it."

Pete said, "Let me clean up this diagram so that we can review what we have done so far. I need a copy of a payment authorization voucher, a budget authorization sheet, a project schedule, a request for bid, and a purchase order."

"I'll get them," answered Sam.

"While you are doing that," said Pete, "think about any other items relating to expenses that cross your desk."

Sam returned with the copies and told Pete that he could not think of any additional items. Figures 5.9 and 5.10 show a payment authorization voucher and a budget authorization sheet, respectively.

Pete showed the reorganized diagram, Figure 5.11, to Sam. "Congratulations," said Pete, "you have just constructed your first data flow diagram." "What?" said Sam.

Data Flow Diagrams

The reorganized diagram of Figure 5.11 shows that engineering department costs are affected by the engineers, the construction project managers, the vendors, and the accounting department.

Figure 5.11 is an example of a **data flow diagram (DFD)**. DFDs are one of the most popular systems analysis modeling techniques in use today. Diagrams drawn early in the process are usually rough sketches done by hand. As the models become more nearly complete, permanent copies are drawn with either a template or a computer.

Rectangles are used to represent entities outside of the system being modeled. These entities are sometimes called **terminators** because they represent the beginning and/or the end of a series of system inputs and outputs.

Payment Authorization Voucher

Vendor: _Jackson Supply_
156 Westridge Ave.
Midwood, NY

PO#: _87-1040_
Contact: _Sam Tilden_
Amount: _324.00_
(See attached invoice for breakdown)

G/L Distribution			Job Cost Distribution	
Account	Amount		Job	Amount
200	324.00			

Department: _Engineering_
Department Manager: _Sam Tilden_
Date: _5/17/88_

FIGURE 5.9

A payment authorization voucher.

Processing activities are represented by circles. In DFD terminology they are simply called **processes**. As the analyst moves through the Model Building Phase of the life cycle, he or she delves deeper into the details of the system's processes.

The paths within the system upon which information and/or objects move are called **data flows** and are represented by arrows.

The DFD in Figure 5.11 shows the data flows that cross the boundary of the cost system to the terminators. For this reason Figure 5.11 is called a **Context Data Flow Diagram** [3] [4]. It illustrates the environment or the context within which the system operates. Drawing the Context DFD is

Budget Authorization Sheet

Department   _Engineering_

Account	Name	Amount Increase/Decrease
100	Engineering	4500.00
200	Materials	1500.00
300	Equipment	1000.00
400	Subcontractors	2500.00
500	Overhead	2000.00

Department Manager: _Sam Tilden_
Project Manager: _Leslie Logan_
Job(s): _C722_

FIGURE 5.10

A sample budget authorization sheet.

the first step of the Model Building Phase of the systems development life cycle.

The Context DFD

In drawing a Context Data Flow Diagram, the systems analyst seeks to

1. establish the boundary of the system under consideration;
2. identify the data flows that cross that boundary;
3. identify the terminators of the data flows in the external environment.

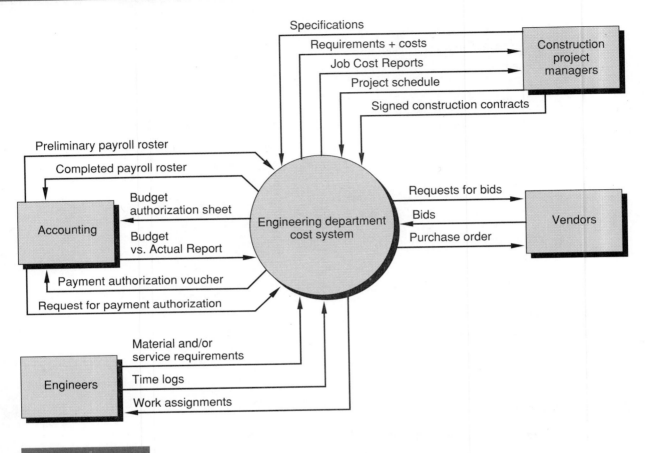

FIGURE 5.11

The Context Data Flow Diagram.

The information contained in the Context DFD will be expanded until a detailed model of the current system is produced.

Since the information in the Context DFD serves as the basis for subsequent steps in the Model Building Phase, it is important. The process of drawing the diagram, however, may be even more important than the finished product because the Context DFD represents the user's first encounter with the technical jargon of the systems analyst. In the Problem and/or Opportunity Analysis Phase the discussion deals with business terms such as objectives, tactics, and efficiency. A knowledgeable manager can see the connection between this material and the problem at hand.

Drawing DFDs, however, can be tedious. It can be a waste of time if the process gets bogged down in semantics and other time-consuming details. The analyst must be diligent in keeping the process productive and in keeping the user motivated and engaged. The best way to do this is to demonstrate continual progress toward the goal of providing a solution to the user's business problem and/or opportunity.

The guidelines for developing the Context DFD shown in Figure 5.12 have been gathered from the encounter between Sam and Pete described earlier.

When first introduced to data flow diagrams, most people try to read them like flow charts. This is wrong. Data flow diagrams do not indicate the sequence of processing steps in a system. They do not show decisions, and they do not show loops in the processing.

A data flow diagram should be read like a road map. The data flows represent the paths or roads connecting the processes of the system. Instead of representing the sequence of processing steps, the DFD shows the paths coming into and going out of each process. Depending upon circumstances, system activity may follow any number of paths from one process to another.

1. Draw the Context DFD with the user(s). Since the objectives of the process are simple, all members of the development team can participate in a single session.

2. Hold the session in a space that encourages free exchange and movement. Use a chalkboard or other medium that allows easy access and modification.

3. Start the session, as Pete did, with a naive representation of the current system. This will introduce the Context DFD symbols and begin the discussion by giving the group something to shoot at.

4. Stick to the goals of establishing the boundary of the current system and identifying terminators and data flows that cross that boundary.

5. Stick to rough sketches to avoid any trauma associated with discarding a diagram and starting over. Plan to throw several diagrams away.

6. Adopt the user's terminology. Refer to documents by their correct names. Make photocopies of all documents and mark them with the names used in everyday practice.

FIGURE 5.12

Guidelines for developing the Context DFD.

The Next Step:
The Analysis DFD

The Context DFD represents the first of three models developed in the Model Building Phase of the systems development life cycle. The second model, the Analysis DFD, is developed through a series of data flow diagrams that represents the processes of the current system in increasing detail.

The first of these diagrams is drawn by separating the data flows of the Context DFD into logical groups, each group representing a system process. For example, the preliminary payroll roster from Accounting, the time logs from the engineers, and the completed payroll roster returned to Accounting form a logical group that might be called COMPILE PAYROLL. The request for payment authorization from Accounting and the payment authorization voucher returned to Accounting make up a logical group that might be labeled AUTHORIZE VENDOR PAYMENTS. Figure 5.13 shows the seventeen data flows of the Context DFD grouped into six different processes. The DFD of Figure 5.13 is called the **Level Zero Analysis DFD** [4].

Choices Regarding the Context and Analysis DFDs

The Context DFD and the Level Zero Analysis DFD are called logical models of the current system. A distinction is usually made between a **logical model** of a system and a **physical model** of a system.

A logical model of a system shows *what* the system does. A physical model shows *how* the work is done. For instance, Sam and Betsy share the work involved in processing payroll and accounts payable transactions, but this physical characteristic is not shown on the Level Zero Analysis DFD. The Level Zero Analysis DFD depicts only *what* is done and not *who* does it.

Pete and Sam are deriving logical models of the current system because the system objectives in Figure 4.2 call for the new system to perform the same tasks as the current system in a better, faster, more flexible way. The new system must also be integrated into the existing Horatio environment, so a thorough understanding of the current system will help with the development of the new system.

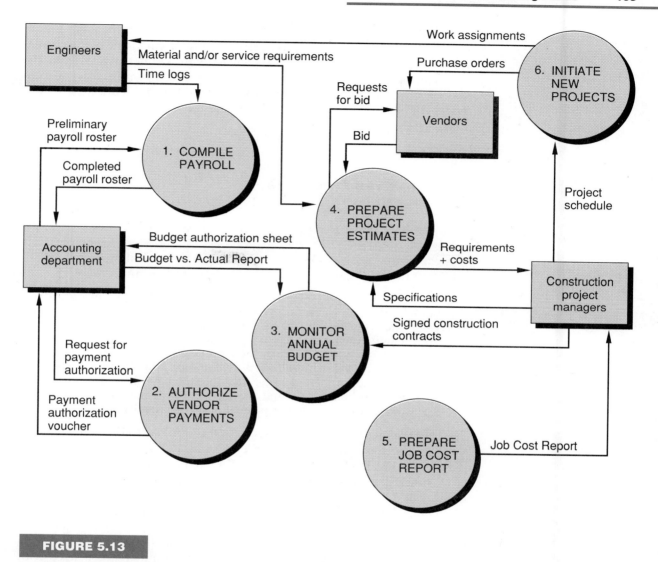

FIGURE 5.13

Level Zero Analysis DFD.

When a new system is needed to do something different from the current system or something new, then the time spent modeling the current system may be wasted, and the development team may choose to model the new system from scratch [6].

Pete was able to begin the Model Building Phase with the logical models because Sam was available to explain the underlying logic of the current system. When experienced users such as Sam are not available to the systems analyst, the analyst observes the current system in order to construct a physical model first. The analyst then tries to deduce the logical model of the current system from the newly constructed physical model [3] [6].

Expanding the Processes

Once the data flows are grouped into the Level Zero Analysis DFD, the analyst expands each of the new processes to increasing levels of detail. This expansion process is known as **leveling**. Let's look at process 1. COMPILE PAYROLL. Recall Sam's description from earlier in this chapter.

> The engineers use time logs to keep strict account of the time spent on each job. Each week Accounting sends me a preliminary payroll roster showing weekly salary for salaried employees and hourly rate for hourly employees. I use the time logs to fill in the hours worked on the roster. I return the completed roster to Accounting and keep a copy for myself. Betsy Klein, our administrative assistant, enters the completed roster data into the payroll system. . . .
>
> Betsy uses the rosters and the logs to break the cost of everybody's time down by job and posts it in the job cost ledger.

These quotes show that process 1. COMPILE PAYROLL is made up of three subprocesses: 1.1 COMPLETE PAYROLL ROSTER, 1.2 ENTER PAYROLL SYSTEM DATA, and 1.3 POST PAYROLL EXPENSE TO JOB COST LEDGER. Figure 5.14 shows the expansion of process 1. COMPILE PAYROLL. This DFD would be called a **Level 1 Data Flow Diagram**.

Notice that the terminators are missing from the level 1 DFD. This is a common practice. Once the Analysis DFD reaches a level of detail that makes the context clear, terminators are left off data flow diagrams to allow more freedom in the sketching. If necessary, the terminator associated with a particular data flow may be found by locating the data flow on a less detailed DFD that shows terminators.

Figure 5.14 introduces a new symbol, the open-ended rectangle. Open-ended rectangles represent **data stores** in DFDs. A data store is any place where information or objects are stored. In this example the job cost ledger, Sam's file for completed payroll rosters and payment authorization vouchers, and the minicomputer-based payroll system are shown as data stores.

When expanding a process, the analyst must **conserve the process's data flows**. This means all data flows that cross the process's boundary on the level 0 diagram must appear in the expanded diagram, and no new data flows that cross the process's boundary can be introduced in the expanded diagram. In Figure 5.13 three data flows—the preliminary payroll roster, the time logs, and the completed payroll roster—cross the boundary of process 1. COMPILE PAYROLL. In Figure 5.14 three data flows—the preliminary payroll roster, the time logs, and the completed payroll roster—cross the boundary. Thus the process's data flows are conserved.

Any new data flows introduced by the expansion of a process must come from or go to data stores. Of course, if the analyst discovers a data flow that crosses a boundary during expansion, then he or she enters that data flow on the level 0 diagram and the expanded diagram as well.

Figures 5.15 through 5.19 represent the expansion of process 2. AUTHORIZE VENDOR PAYMENTS through process 6. INITIATE NEW PROJECTS. The information contained in the dialogue between Sam and Pete presented earlier was used to carry out the expansion. Study these figures, taking note of the connection with the dialogue, the conservation of data flows, and the introduction of additional data stores.

Pay particular attention to the expansion of process 5. PREPARE JOB COST REPORT. After expansion, no process may be left with only inputs or only outputs. In the level 0 diagram, Figure 5.13, process 5 had only the Job Cost Report output data flow associated with it. The expansion of the process revealed that the job cost ledger data store is the source of the information contained in the Job Cost Report.

The six expansion DFDs, Figures 5.15 through 5.19, make up the level 1 diagrams for the cost system. Since the cost system is straightforward, no further leveling is needed. The analyst stops the leveling process when all members of the development team feel they understand the current system. If subsequent diagrams are drawn, they are labeled **Level 2, Level 3,** etc.

If the system processes are very complicated, the analyst may resort to a written description after the expansion DFDs reach a certain level. **Structured English** is a popular tool for this purpose. It relies upon the look and feel of computer programming languages to produce clear and concise process descriptions. Examples of structured English process descriptions are presented in Chapter 13.

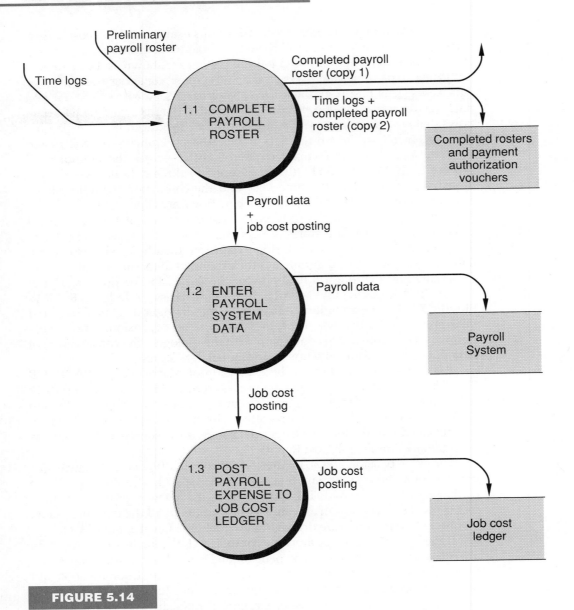

FIGURE 5.14

Level 1 of process 1. COMPILE PAYROLL.

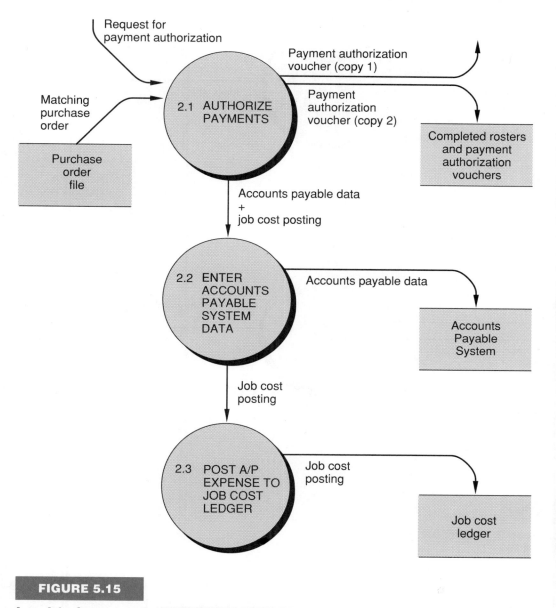

FIGURE 5.15

Level 1 of process 2. **AUTHORIZE VENDOR PAYMENTS.**

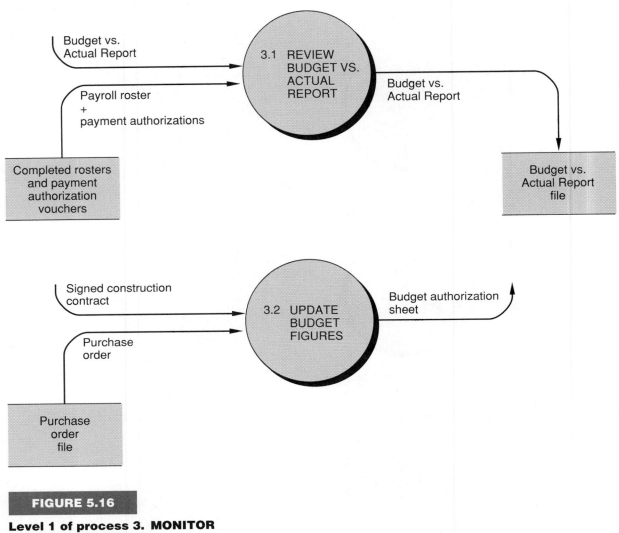

FIGURE 5.16

**Level 1 of process 3. MONITOR
ANNUAL BUDGET.**

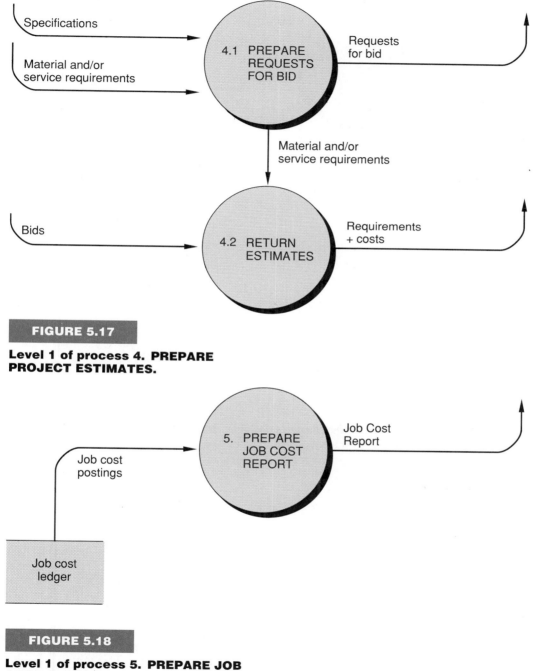

FIGURE 5.17

Level 1 of process 4. PREPARE PROJECT ESTIMATES.

FIGURE 5.18

Level 1 of process 5. PREPARE JOB COST REPORT.

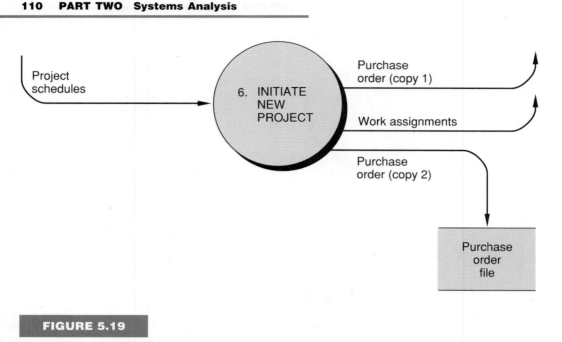

FIGURE 5.19

Level 1 of process 6. INITIATE NEW PROJECTS.

Connecting the Processes

The final version of the Analysis DFD combines the data stores uncovered in the leveling process with the Level Zero Analysis DFD. In the final version data stores serve to connect processes by way of data flows. Figure 5.20 lists the processes of the Level Zero Analysis DFD and the data stores uncovered during leveling.

Notice that the job cost ledger, the completed rosters and payment authorization vouchers, and the purchase order file data stores are each associated with three processes. These are the central data stores. They should occupy the center of the final version of the Analysis DFD, Figure 5.21.

Processes 1. COMPILE PAYROLL and 2. AUTHORIZE VENDOR PAYMENTS both access the job cost ledger and the completed rosters and payment authorization vouchers, so these processes should be located near these data stores, Figure 5.22.

Process 5. PREPARE JOB COST REPORT accesses only the job cost ledger data store, so it should occupy a corner position. The same is true for

Process	Data Store
1. COMPILE PAYROLL	Job cost ledger Payroll System Completed rosters and payment authorization vouchers
2. AUTHORIZE VENDOR PAYMENTS	Purchase order file Job cost ledger Accounts Payable System Completed rosters and payment authorization vouchers
3. MONITOR ANNUAL BUDGET	Budget vs. Actual Report file Completed rosters and payment authorization vouchers Purchase order file
4. PREPARE PROJECT ESTIMATES	
5. PREPARE JOB COST REPORT	Job cost ledger
6. INITIATE NEW PROJECTS	Purchase order file

FIGURE 5.20

Cost system processes and data stores.

process 4. PREPARE PROJECT ESTIMATES, which accesses no data stores. The remaining data stores are fitted into the Final Analysis DFD as efficiently as possible, Figure 5.23. This usually takes several attempts, so rough sketches should be used in the early stages.

In the **Final Analysis DFD**, Figure 5.23, the analyst records on one diagram the data flows of the Context DFD (there were 17 of these), the processes of the Level Zero Analysis DFD (there were 6 of these), and the data flows and the data stores of the expansion process. It is difficult to keep the data flows from crossing, but the analyst should strive for this through the proper positioning of the data stores and processes.

Sometimes an active data store must be repeated in order to produce a readable diagram. In such a case the analyst marks the duplicate data store symbols so that the reader knows the data store appears more than once on the diagram.

Figure 5.24 illustrates this technique using asterisks. Vertical hash marks are also used to mark duplicate symbols. Try drawing Figure 5.24 without the duplicate data store symbols, and judge the readability of the diagram for yourself.

Job cost ledger	Completed rosters and payment authorization vouchers	Purchase order file

FIGURE 5.21

The central data stores of the Analysis DFD.

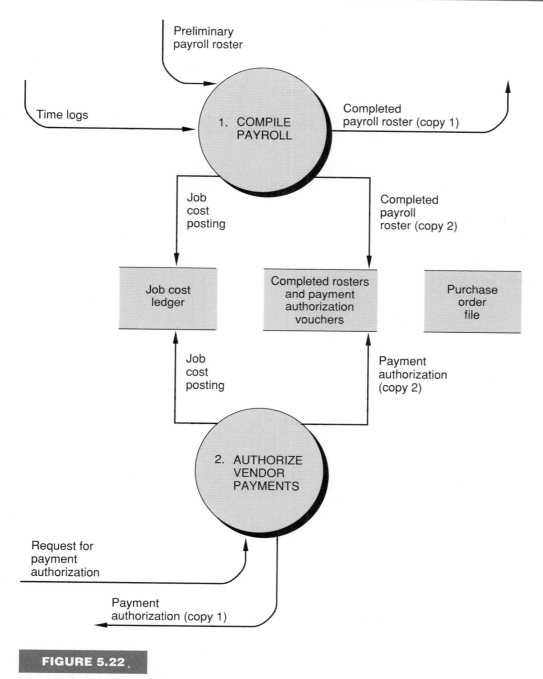

Locating processes on the Analysis DFD.

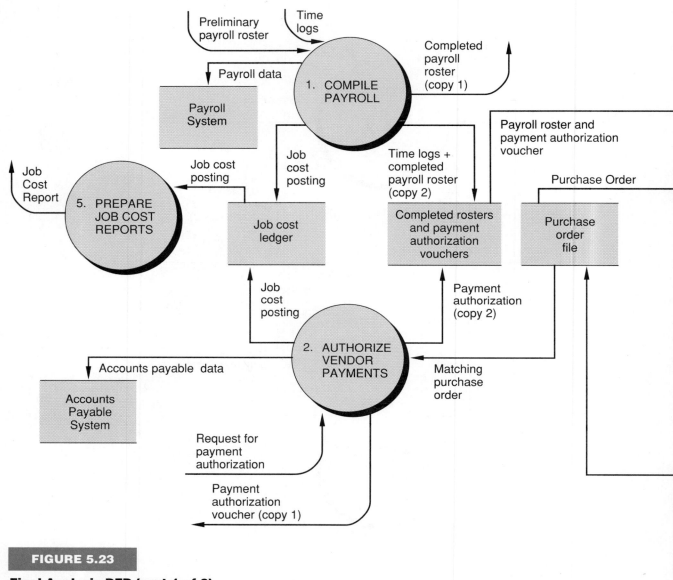

FIGURE 5.23

Final Analysis DFD (part 1 of 2).

Summary of the Analysis DFD

Figure 5.25 summarizes the steps involved in the construction of the Analysis DFD, the second model of the Model Building Phase of the systems development life cycle.

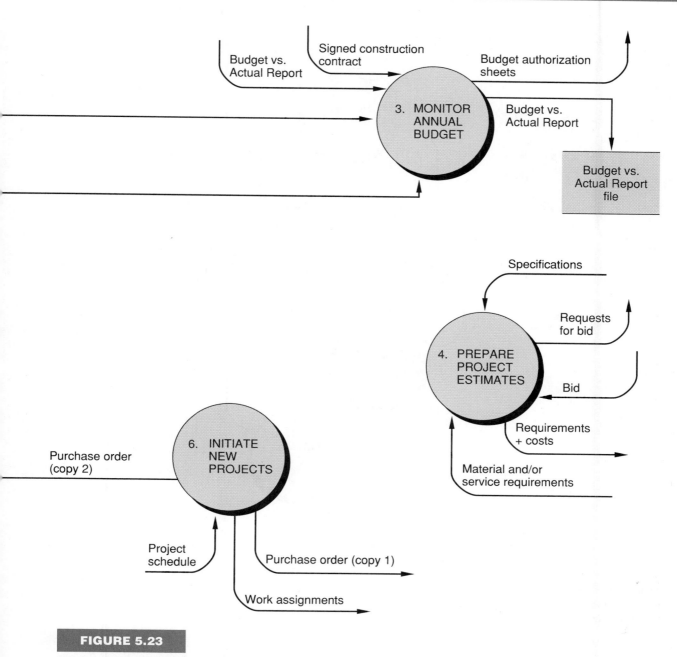

FIGURE 5.23

Final Analysis DFD (part 2 of 2).

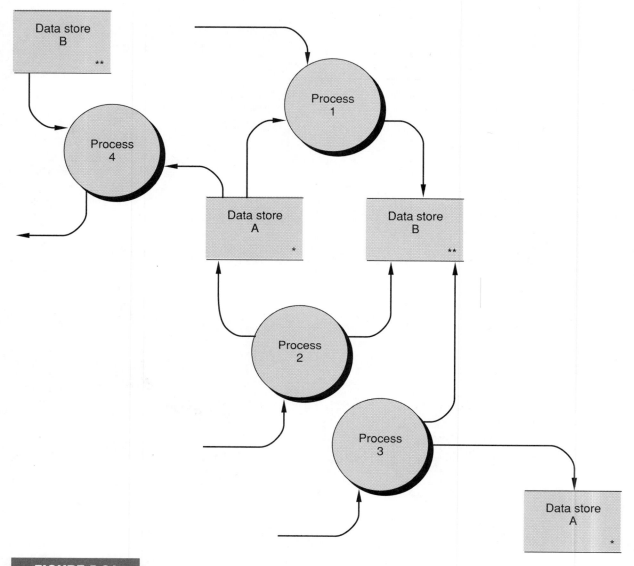

FIGURE 5.24

Using duplicate data store symbols to
avoid intersecting data flows.

1. Draw the Level Zero Analysis DFD by separating the data flows of the Context DFD into logical groups. Connect each group of data flows to a process. Name each process with a transitive verb and a direct object.

2. Expand each of the level 0 processes into subprocesses. Conserve the level 0 data flows and use data stores to introduce new data flows. Continue expanding the subprocesses until an adequate level of detail is achieved.

3. Combine the data stores and the Level Zero Analysis DFD into the final version of the Analysis DFD. Observe the connections between processes established by the inclusion of the data stores in the diagram.

FIGURE 5.25

Steps in the construction of the Analysis DFD.

The Requirements Model

With the completion of the Analysis DFD, Sam and Pete broke for the day. They agreed to meet two days later to allow time to absorb the developments of the day. Pete advised Sam to review the plan he submitted to Mr. Chapin, Figure 4.2, and to imagine how his work would change if a new system were in place. Before they left the room, Sam wrote DO NOT ERASE in large letters on the chalkboard.

Pete began the second meeting. "Today we will compare the plan you submitted to the president with the Analysis DFD in order to develop the **Requirements Model** for the new system. The Analysis DFD presents a picture of the cost processing system of the engineering department. Can any process or context data flow be eliminated because it serves no useful purpose?"

"I don't think so," replied Sam. "If we read clockwise around the diagram, you're talking about the Job Cost Report, payroll rosters, budget authorization sheets, requests for bid, costed requirements, purchase orders, and payment authorization vouchers. I don't see how we can eliminate any of those. They are all part of my job responsibility. Maybe the budget authorization sheets can go if we set up our own monthly system."

"We'll have to see," said Pete. "Your data is probably part of several company-wide reports, so it is not likely you'll be able to remove yourself completely from the Accounting Department System. It's probably not a good political move either."

"You're right," said Sam.

"So," said Pete. "We know, then, that the new cost system we design must provide all of these functions and more."

"We're going to put all these processes on the computer?" asked Sam.

"Not necessarily," said Pete. "We'll decide that later. At this point the term *system* refers to both manual and automated processes." Pete continued, "Let's look at the plan you submitted to the president."

Sam took the plan, Figure 4.2, from his file. Pete asked, "What is the first tactic?"

"Eliminate manual preparation of job cost reports," said Sam.

"Is all the data for the Job Cost Reports in the data stores of the Analysis DFD?" asked Pete.

"I think so," said Sam. "All the information we need is in the job cost ledger. The problem with the ledger is the time it takes to enter the information, the errors made in copying the data, and the effort required to prepare special reports manually."

"OK," said Pete. "We've got the data for the Job Cost Reports covered, but we need to add a convenient way to access this data. What's the next tactic?"

Sam checked his plan. "Institute a monthly budgeting system by job number and general ledger account to replace the current annual system. We do not have the data or the access mechanism for a monthly budgeting system on the Analysis DFD." Sam continued, "Reorganize the vendor base for higher volumes with fewer vendors is the next tactic. The data to perform the analysis is in the completed rosters and payment authorization vouchers data store on the Analysis DFD, but we do not have convenient access to this data. Also, we need a way to control purchase orders after the reorganization is done. We tried this once before. We did a lot of work to identify our key vendors, but after awhile, we were back to the same hit-and-miss game."

Pete made some entries on his notepad. "Is that it for the tactics?"

"No," said Sam. "You forgot the last one. Institute professional development plans for engineering personnel. We've got the data for that in the completed rosters and payment authorization vouchers data store, but we need a way to set up the plans and a way to monitor progress. That's it for the tactics."

"Fine," said Pete. "Let's look at the system objectives from the plan."

Sam read, "Budget vs. Actual Reports on a monthly basis, Job Cost Summary and Detail Reports, Detail INQUIRY BY SOURCE of expense or vendor."

Pete said, "We've already noted what we need to achieve these objectives. Can you think of other requirements to add to the list?"

"Are you kidding?" said Sam. "My head is spinning; I would not trust anything I came up with now. I know we've got a lot down on that list, but

I can't be sure the list is complete. I'd be uncomfortable signing off on it now."

"You don't have to," answered Pete. "Let's go with this for the moment, and we'll see if it leads to anything else." Pete organized his notes into the list shown in Figure 5.26.

Guidelines for the Requirements Model

The Requirements Model is the third, and final, model developed in the Model Building Phase of the systems development life cycle. Development guidelines are listed in Figure 5.27.

Summary

Problem and/or Opportunity Analysis and Model Building are the two most difficult phases of the systems development life cycle. Errors and omissions in these phases lead to solving the wrong problem. No matter how elegant or sophisticated the solution, it is useless if it does not address

Cost System Requirements
1. Continue to perform the functions and maintain the data of the Analysis DFD.
2. Provide a convenient access mechanism for the data of the Job Cost Detail and Summary Reports.
3. Provide for the maintenance and use of monthly budget data by job number and general ledger account.
4. Provide the analysis required for the reorganization of the vendor base. Provide a control mechanism for choice of vendor.
5. Provide for the maintenance and use of professional development plan data for engineering department personnel.

FIGURE 5.26

The cost system requirements model.

1. Examine the processes and context data flows of the Analysis DFD to determine if any no longer serve a useful purpose.

2. Examine each business tactic developed in the Problem and/or Opportunity Phase. Determine the availability of required data in the Analysis DFD data stores. Be concerned only with the existence of the data; do not be concerned with the means by which it is stored.

3. For each business tactic, judge the adequacy of access and control mechanisms.

4. Examine each system objective developed in the Problem and/or Opportunity Analysis Phase for new system requirements regarding data and processes.

FIGURE 5.27

Steps in the construction of the Requirements Model.

the problem at hand. The talents listed on the first page of this text must all be brought to bear upon the tasks of these two critical phases.

The Horatio & Co. Cost Control System is a good first experience in problem analysis and model building because it is realistic and simple. In working the exercises, follow the checklists presented in this chapter. As you gain experience, you will develop your own style of proceeding through these tasks.

You should be aware that there are several data flow diagramming styles [3], [5] and that data flow diagramming is only one of many modeling techniques in use today. The book by Aktas [1] contains short chapters on many of these other techniques. Follow the references in that book if you wish to study a particular technique more deeply.

THOUGHT QUESTIONS
In Your Opinion. . .

1. What similarities and differences do you see in developing a system to replace an existing system and developing a system to do brand new work? Give an example of each type of development project. On which would you rather work? Why?

2. Imagine you are about to begin your first assignment as a systems analyst. Your project manager says to you, "We never model the current system here. We consider the needs of the business, and we decide what the new system should do." What is your reaction?

3. Do you think the amount of time and effort required to draw and redraw data flow diagrams discourages people from using them? Is this a valid reason for not using data flow diagram models? Now that modeling systems are available for microcomputers, do you think we will see an increase in the use of DFDs?

4. What do you think Pete meant when he said that setting up an independent monthly budgeting system for the Engineering Department was not a good political move? How important do you think political factors are in systems development?

5. What did you think of the way Pete ran the modeling sessions with Sam? Does Pete have a pattern to the way he runs these sessions? If yes, what are some of the techniques he uses? Do you think the way in which a modeling session is run affects the final models? the final system? Why or why not?

6. What did you think of Sam's participation in the modeling sessions? Do you think the interaction between Sam and Pete is typical of system development projects? Why or why not?

References

1. Aktas, A. Z. *Structured Analysis and Design of Information Systems*. Englewood Cliffs, NJ: Prentice-Hall, Inc., 1987.

2. Bostrom, R. "Successful Application of Communication Techniques to Improve the Systems Development Process," Indiana University Working Paper, 1987.

3. DeMarco, T. *Structured Analysis and System Specification*. New York: Yourdon Press, 1978.

4. Eckols, S. *How to Design and Develop Business Systems*. Fresno, CA: Mike Murach and Associates, Inc., 1983.

5. Gane, C., and T. Sarson. *Structured Systems Analysis Tools & Techniques*. New York: Improved System Technologies, 1977.

6. Gane, C. *Rapid System Development*. New York: Rapid System Development, Inc., 1987.

HORATIO&CO.

Credit Union Processes

In this section you will develop the Context DFD, the Analysis DFD, and the Requirements Model for the Horatio & Co. Credit Union System. Since you cannot work with Camille Abelardo, use the information given in the assignment section of Chapter 4, your answers to the Chapter 4 assignments, and the following information to develop the models.

Deposits and withdrawals are carried out as they are in a bank. The member completes a two-part deposit or withdrawal form. The teller retrieves the member's account card, enters the transaction on the card, initials the deposit or withdrawal form, stamps it with the date, returns the copy to the member, enters the transaction on the daily activity log, and files the original deposit or withdrawal form with the log.

Camille collects the logs and the forms at the end of the day to do the cash reconciliation. If the transaction is a withdrawal, the teller must also verify the member's signature against a file of signature cards kept in alphabetical order near the teller's window. Deposit and withdrawal forms are shown in Figures 5.1A and 5.2A, respectively.

DEPOSIT — HORATIO & CO.
EMPLOYEE CREDIT UNION
DATE: 12/10/86

ACCOUNT: K101
NAME: Mary Kilpatrick
ADDRESS: 4 East 88TH Street
New York, New York

CASH →		
Check No....	25	00
102	65	00
435	10	00
TOTAL	100	00

FIGURE 5.1A

A sample deposit form.

WITHDRAWAL — HORATIO & CO.
EMPLOYEE CREDIT UNION

ACCOUNT: B313

NAME: Erika Berg

ADDRESS:58 Victory Place

x   Nutley, NJ

x

DATE: 12/15/86

RECEIVED FROM HORATIO & CO. CREDIT UNION 200.00

Two Hundred ——————————————————— DOLLARS

Erika Berg

MEMBER'S SIGNATURE

WDR

TELLER'S INITIALS

FIGURE 5.2A

A sample withdrawal form.

To enroll as a new member, the employee completes two signature cards. The signature card contains the member's name and address and the conditions of membership in the credit union. A credit union representative determines the next available account number and writes it on both cards. One card is entered into the alphabetical card file, and the member keeps the other card for reference. In addition to verifying signatures on withdrawals, the alphabetical card file is used to look up account numbers for members who have forgotten their number.

Exercises

1. Based upon your own experiences with credit unions, banks, and other institutions, draw a naive Context DFD for the manual credit union system.

2. Use the section entitled *The Credit Union's Problems and Opportunities* from the assignment section of Chapter 4 and the above information to develop the Context DFD for the manual credit union system.

3. Use the section entitled *The Credit Union's Problems and Opportunities* from the assignment section of Chapter 4, the above information, and your work in exercise 2 to develop the Level Zero Analysis DFD for the manual credit union system.

4. Use the section entitled *The Credit Union's Problems and Opportunities* from the assignment section of Chapter 4, the above information, and your work in exercise 3 to develop the Level 1 Analysis DFDs for the manual credit union system.

5. Use the section entitled *The Credit Union's Problems and Opportunities* from the assignment section of Chapter 4, the above information, and your work in exercise 4 to level the DFDs further, if necessary, and to develop the Final Analysis DFD for the manual credit union system.

6. Use the section entitled *Camille's Problem and/or Opportunity* from the assignment section of Chapter 4 and your work in the Chapter 4 exercises and the exercises above to develop the Requirements Model for the new credit union system.

CHAPTER 6

Developing Solution Alternatives

Objectives

The Analysis DFD, Figure 5.23, represents a logical model of the current system. Logical models of the current system are developed whenever knowledge of the current system contributes to the development of new system requirements. If experienced users are not available to describe the underlying logic of the current system, then the logical model is deduced from a physical model of the current system which is developed through observation of current system operations.

The Requirements Model, Figure 5.26, represents the goal of the project. It does not, however, represent a system. In this chapter Pete and Sam transform the Requirements Model into a logical model of the new system. This model is called the **Design DFD**. It represents the processes, data stores, and data flows that the new system must provide to satisfy the Requirements Model. The development of the Design DFD represents the start of the Evaluation of Alternatives Phase of the systems development life cycle.

Chapter 7 describes the process by which the Design DFD is transformed into a model of a feasible physical form. This transformation from the Design DFD to a physical model of the new system completes the Evaluation of Alternatives Phase of the life cycle.

The underlying logic of the life cycle's analysis phases which begin with current system models and end with new system models [2] is summarized in Figure 6.1.

The specific objectives of this chapter are

1. To determine the boundary of the new system, i.e., the work that must be performed to satisfy the Requirements Model;

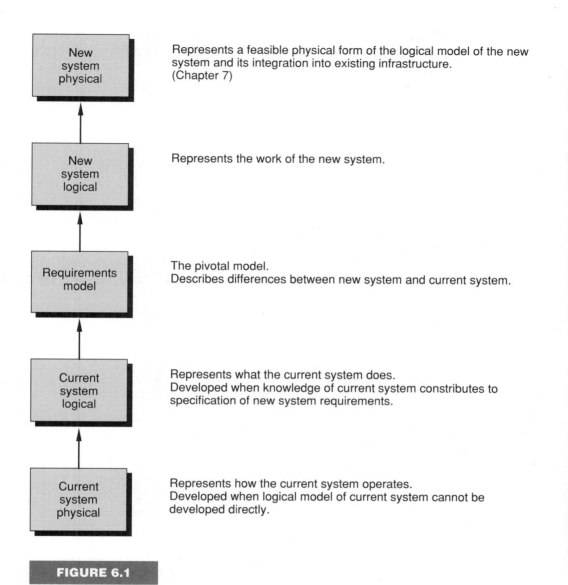

New system physical	Represents a feasible physical form of the logical model of the new system and its integration into existing infrastructure. (Chapter 7)
New system logical	Represents the work of the new system.
Requirements model	The pivotal model. Describes differences between new system and current system.
Current system logical	Represents what the current system does. Developed when knowledge of current system constributes to specification of new system requirements.
Current system physical	Represents how the current system operates. Developed when logical model of current system cannot be developed directly.

FIGURE 6.1

A logical model of the Analysis Phases of the systems development life cycle.

2. To identify the work of the new system best suited to automation through an analysis of benefits, the likelihood these benefits will occur, and the form these benefits will take;

3. To identify the critical data stores of the new system, the relationships among the data stores, and the potential sources of the contents of the data stores;

4. To combine the work described above into the Design DFD, a logical model representation of the new system.

Understanding the New Processes

Sam copied the Cost System Requirements Model (shown again in Figure 6.2) onto the chalkboard alongside the final version of the Analysis DFD.

"Our Requirements Model specifies that the new system will do everything that the current system does and more," said Pete. "Today we'll concentrate on understanding more about the new work, the processes specified in the Requirements Model that do not appear in the Analysis DFD. That would be number 3—Monthly Budgets, number 4—Vendor Control, and number 5—Professional Development Plans."

COST SYSTEM REQUIREMENTS

1. Continue to perform the functions and maintain the data of the Analysis DFD.

2. Provide a convenient access mechanism for the data of the Job Cost Detail and Summary Reports.

3. Provide for the maintenance and use of monthly budget data by job number and general ledger account.

4. Provide the analysis required for the reorganization of the vendor base. Provide a control mechanism for choice of vendor.

5. Provide for the maintenance and use of professional development plan data for engineering department personnel.

FIGURE 6.2

The Cost System Requirements Model (repeated from Figure 5.26).

Professional Development Plans

"Let's start with number 5—the Professional Development Plans," said Sam. "I have been doing this on an informal basis already. Each engineer meets with me twice per year for a performance review. The results of the fall meeting determine salary increases that take effect the following January. We use the spring meeting to touch base and to mark progress toward goals established the prior fall."

"How would you run these meetings if you had the new system?" asked Pete.

"At these meetings I've started to make commitments to the engineers to provide certain kinds of work," said Sam. "Design work is the most desirable, but not everyone has the experience to qualify for it. Those who do not have the experience expect to develop it gradually. If I do not provide the means to develop an engineer's skills, he or she will leave the company. It's a seller's market out there, so it's fairly easy to find a good job. The same competition that's forcing us to cut costs is creating pressure in the job market for engineers.

"With the new system I could go to the review meetings with an up-to-the-minute, accurate tabulation of the engineer's activity for the past six or twelve months. This way we could see exactly how the engineer spent his or her time and make better plans for the future."

"Where is that information now?" asked Pete.

"The information is in the time logs," said Sam. "I tried compiling it before the meetings, but I see two engineers per day during review time. Eventually, I gave up. In addition, I need the tabulations updated every week, so that I can review them when I make new assignments."

"What about asking Betsy or someone else to compile reports like that?" asked Pete.

"I tried that too. The turnaround time was still a problem," answered Sam. "Besides, when someone else does them, I still have to review them for accuracy. Also, the tabulations require some engineering judgment. If I don't do them, I get interrupted so many times with questions that I don't get anything else done anyway. With the new system, I'm hoping to be able to call this information up on the screen with the touch of a button, use it, and get rid of it. If I need it again, I'll call it up again."

"OK," said Pete. "I'm getting the picture. If you make commitments to the engineers, you must have some idea of what kind of work is coming up. Where do you get that information?"

"I always have a fair idea of what's coming up," said Sam. "Before I sit down with the engineers, I review the signed contracts to bring myself up to date on the details. These are the same documents we use to prepare the annual budget authorization sheets. Of course, nothing is etched in stone at the review meetings. If situations change, then assignments change. Everybody understands that."

"Let's recap what we have done so far," said Pete. "Try using a Level 1 DFD to express how you would handle professional development plans with the new system in place. We had six processes on the Analysis DFD, so number this one process 7."

Sam sketched Figure 6.3 on the chalkboard. "The critical thing for me is the Activity Summary data flow going into 7.2," said Sam. "That has to be effortless and fast. The Professional Development Plan data store does not have to be automated. The plan usually amounts to a half-page of notes that I keep in the engineer's personnel file."

"Do you think the process is accurately represented in the diagram?" asked Pete.

"Yes, it is," said Sam. "Are we going to sketch one of these for every new process in the Requirements Model?"

"Yes," said Pete. "This technique has always worked for me. Try another of the new requirements."

Monthly Budgets

Sam reviewed the Requirements Model on the chalkboard. "Number 3 is fairly straightforward," he said. "The format of the current Budget vs. Actual Report is fine for the financial statements. The problem is that the annual budget figures do not identify overruns fast enough and do not provide breakdowns by job. I try to estimate monthly expenditures by job, but I get lost in the details. A lot of problems slip by. We need to deal with Monthly Budgets by Job in a systematic way.

"I would say the processing for monthly budgeting would be the same as for annual budgeting," continued Sam. "The difference would be the use of Monthly Budget figures by both Job and general ledger Account instead of annual budget figures by general ledger account alone."

Pete made a photocopy of the Level 1 DFD for process 3. MONITOR ANNUAL BUDGET of the Analysis Data Flow Diagram, Figure 5.16. He changed the caption to read Level 1 of process 8. MAINTAIN MONTHLY BUDGETS. Sam was satisfied that model accurately represented how he

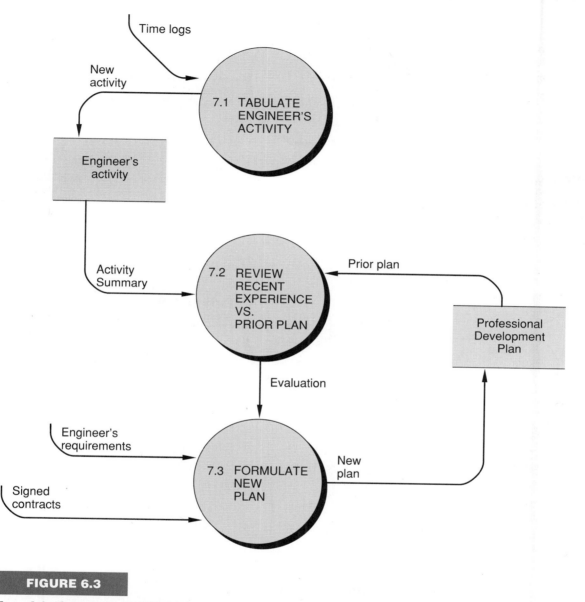

Level 1 of process 7. MAINTAIN PROFESSIONAL DEVELOPMENT PLANS.

would handle Monthly Budgets if the new system were in place. The diagram is shown in Figure 6.4.

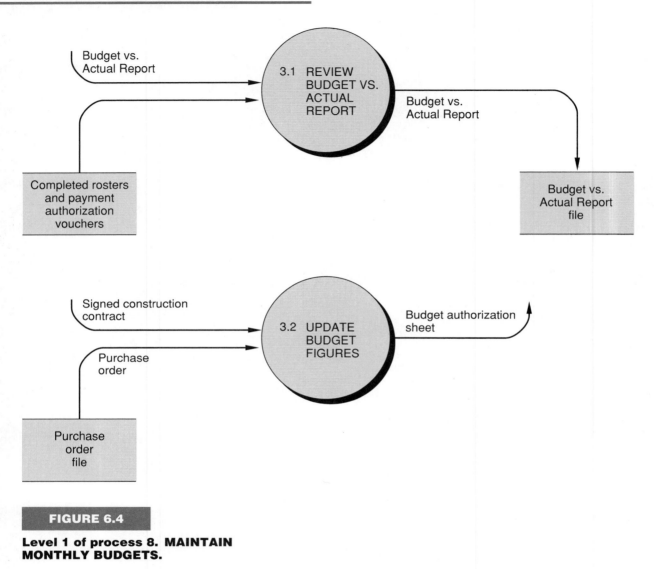

FIGURE 6.4

Level 1 of process 8. MAINTAIN
MONTHLY BUDGETS.

Reorganizing the Vendor Base

"What about requirement number 4—Reorganize the Vendor Base?" asked Pete.

"I have been avoiding that one," said Sam. "I know what I want to achieve, but I don't have a clear idea of how to do it, and I don't even know if it's worth doing."

"Just think out loud for awhile, and try to avoid making value judgments of new ideas at this point," said Pete.

"Simply stated, I want to lower our costs on materials, equipment, and subcontractors. Over the years we have identified the cheapest source of supply for almost every item we buy. We tend to focus on the individual item costs, and lose sight of the big picture. Some vendors would be happy to put together a complete package, but they do not even get a chance to bid on all the items they carry because we have fallen into the habit of going to certain vendors for certain things. I think we are spread too thin.

"The company is much better organized on the construction side. I've discussed this a few times with Joe Sindora, the construction manager. He does not use a computer system, but after 30 years of experience, he has become a walking encyclopedia of who carries what products and who puts together the best packages."

Pete walked to the chalkboard and began to sketch. His diagram is shown in Figure 6.5. "What do you say to the development of a vendor control data store that contains things like which vendors want to bid on what products? You would use the data to identify appropriate vendors when you are preparing requests for bid. There would be a one-time effort to get started, then you would keep the data current through periodic updates that either you or the vendors initiate."

"That sounds just like the vendor lists that the government agencies maintain," said Sam. "We're on several of them. I never thought of the situation in quite that way, but that is exactly what Joe does, except he does not use a computer."

With the completion of Figure 6.5, Pete suggested a break for lunch. "We'll discuss the value of automating some or all of the system processes when we return," he said.

Summary

In the previous section Pete and Sam focused on the new system processes not shown on the Analysis DFD for the current system. Together, they expanded their understanding of these processes from the few sentences of the Requirements Model to the level 1 data flow diagrams of Figures 6.3, 6.4, and 6.5.

The goal of the first step of the Evaluation of Alternatives Phase is simply to begin to understand the details of the new processes specified in the Requirements Model. The leveled DFD provides a convenient way to develop such understanding. In the previous section no attempt was made

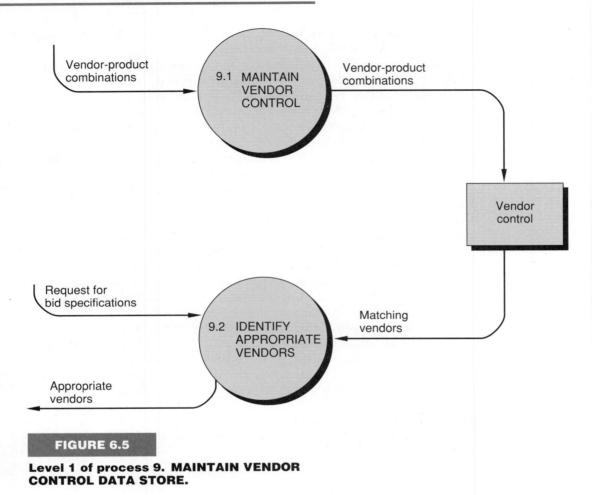

FIGURE 6.5

**Level 1 of process 9. MAINTAIN VENDOR
CONTROL DATA STORE.**

to integrate the new requirements into the Analysis DFD. The Analysis DFD represents the current system, and the new processes do not exist in the current system.

The processes of the Analysis DFD included in the Requirements Model and the new processes specified in the Requirements Model represent the **boundary of the new system**, i.e., the work the new system must do. In the steps to follow, Sam and Pete will try to integrate these processes into a working system. They will make choices depending upon the value of each alternative.

The Value of Automation— Job Cost Reports

Pete began the afternoon session. "The processes of the Analysis DFD included in the Requirements Model and the new processes specified in the Requirements Model represent a first draft of the boundary of the new system," he said. "This is the work we think the new system will do. Now we need to examine how automation fits into the picture. We need to determine what the computer will do and what Horatio & Co. personnel will do.

"We want to look at the items in our Requirements Model, Figure 6.2, in terms of value," continued Pete. "Generally speaking, value involves system benefits and costs. Let's concentrate on benefits without regard to costs for the moment. Through which requirement will automation deliver the greatest benefits?"

"That's easy," said Sam. "Number 2—Job Cost Reports. The manual system does an adequate job for annual accounting purposes, but we have to hire someone full-time, usually an evening college student, to prepare other reports. It is not exciting work, and there is a lot of pressure in the job. Even though we pay $18,000 per year, there is a lot of turnover in the position. The turnover creates real problems; the work backs up, and we have to spend time finding and training a replacement.

"The biggest problem with job cost information, however, is that it often holds up the work of 20 very expensive engineers and technicians. The company loses in terms of morale as well as dollars."

"How does the report preparer spend his or her time?" asked Pete.

"Right now, it's her time," said Sam. "The student's name is Kim Long. I introduced you to her the last time you were here."

"I remember," said Pete.

"Kim spends most of her time compiling job cost ledger data into reports," said Sam. "That involves collecting data, organizing it, doing calculations, typing, proofreading, and photocopying. The end-of-the-month Job Cost Reports to the project managers are fairly simple, and since we know about them in advance, we can schedule Kim's time to prepare them. The random requests that occur during the month create much more difficulty."

"Give me an example," said Pete.

"OK," said Sam. "Rita Griel is one of our engineers. She called me this morning to ask how much we spent on Engineering for Job C722; she was preparing an estimate for a similar job. Kim is working on another project, so the work on Rita's request will not even begin until tomorrow afternoon. The Budget vs. Actual Report tells how much we spent for Engineering for all jobs, and the Job Cost Report tells how much we spent on Job C722 for all accounts. To answer the question, Kim will have to plow through each line of the Job Cost Ledger to collect all the postings for Engineering, type up a report, photocopy it, send a copy to Rita, and file one for me."

"Why do you keep a copy?" asked Pete.

"I keep a copy to avoid duplicate effort," said Sam. "Before Kim compiles one of these reports from scratch, she checks the file to see if we have done something like it before. Sometimes we get away with compiling a small portion and cutting and pasting the rest from previous reports. It's a real nightmare."

Sam continued, "Even if the request does not involve jobs, we usually go to the job cost ledger to compile the data."

"Why is that?" asked Pete.

"Convenience," said Sam. "All the data is in the one book, and it's all in the same format. You saw the job cost ledger page. We keep date, source, account, and amount for each expense; all expenses on a particular page are for the job listed at the top of the page. Because each item is always in the same place, it's much easier to look through the ledger than to fish through the payroll rosters or the payment authorization forms."

"What about accuracy?" asked Pete.

"We have serious problems," said Sam. "The transcription and calculation errors are bad enough, but sometimes we discover errors in the postings to the ledger. That shakes everybody's confidence in the whole process."

"You've made quite a strong case," said Pete. "Let's try to quantify some of these benefits."

"Well," said Sam, "there is the $18,000 per year for Kim, which does not include fringe benefits. As manager, I've never let anyone go, but this situation is intolerable as it is. We always need people like Kim at Horatio. If she wants to stay with the company, we'll find another place for her.

"More importantly, there is the time spent by a staff of 20 engineers and technicians doing reports themselves, waiting for reports, looking for files that are being used by someone else, and tracking down errors in the reports. With the turnover in Kim's position, I would estimate that each person spends or wastes an average of one hour per day on the manual Job Cost Reports."

"That's one-eighth or 12.5% of the workday," said Pete. "What is the average salary for engineers and technicians?"

"$50,000 including fringe benefits," said Sam.

"So the manual preparation of Job Cost Reports costs the company $50,000 times 12.5% times 20 plus $18,000 per year," said Pete.

"Yes," said Sam. "$143,000 per year for a product that is making everyone miserable."

The Value of Automation— Turnover

"If each engineer and technician saves one hour per day, what will be done with the time?" asked Pete. "Eliminating the manual preparation of Job Cost Reports would provide such a boost in morale that the company would recover more than one hour per day from each person," said Sam. "Since our revenues are expected to decrease in the future, I would look to shrink the size of our staff."

"I thought you were worried about turnover," said Pete.

"I am," answered Sam. "This will help the turnover situation a great deal. It's true that the elimination of the clerical work will cut the number of positions in the department, but the remaining positions will be of higher quality, which will serve to improve stability."

"Can you quantify the cost of turnover?" asked Pete.

"That's hard," said Sam. "We've got the employment agency fee, which is about $3,000. The new person is almost always more expensive; our last three replacements cost us an extra $3,000 each. Then we've got our own time in reassigning work, reviewing resumes, and interviewing, which comes to about three person-days. At $50,000 per year, that's worth about $600. But these direct expenses represent only half the cost picture.

"A new person is not fully productive until he or she has been here for a month. One would expect this, but there is a hidden cost factor in turnover that I am just coming to realize. Everyone on the project team to which the new person is assigned is less productive until the new person is assimilated into the project. Each engineer must establish communication with the new person, which takes time, which delays the project work further. As the schedule slips, tensions rise, making people even less productive. It's a vicious cycle. I cannot estimate the cost of this phenomenon. A conservative figure would be an average engineer's salary for one month, which does not take into account the effect of our delays on other departments."

"That's $50,000 divided by 12, or roughly $4,200," said Pete.

"That's low, but let's go with it," said Sam. "So we have a total of $10,800 as a conservative estimate of the cost of replacing an engineer," said Pete. "How often do you have to replace an engineer?"

"There were three replacements last year," said Sam. "If we do nothing, we could have the same situation again this year."

"What are the chances that an automated system will prevent someone from leaving?" asked Pete.

"By itself, no chance," said Sam. "The automated system will allow me to understand the operation of this department better. With this understanding I will make better decisions regarding assignments. It's what I do with the information that will make a difference.

"Right now, I am in the dark. I make assignments based upon what I remember of the plans and recent projects. I confess that sometimes I respond to the person who complains the most."

"What about a separate file of assignments?" asked Pete.

"Of course, I keep track of who is assigned to what project," said Sam. "But the format of that information is not conducive to tracking an individual's professional development."

Pete said, "I think we have covered quite a bit of ground. Let's meet again first thing in the morning."

"I can't," said Sam. "I'm going out to a construction site in the morning. How about after lunch?"

"Fine," said Pete.

Summary

Recall Sam's business plan presented in Figure 4.2. In this plan Sam listed four business tactics for achieving the business objective of maintaining the current level of profit. In developing the Requirements Model, Figure 6.2, Sam and Pete expressed the business tactics of Sam's plan in terms of data flow diagram data stores and processes.

Requirements Model processes that were not part of the Analysis Data Flow Diagram were expanded to the same level as those of the Analysis DFD, and Pete identified the combined set of processes as the boundary of the new system.

In the preceding section Sam and Pete began to explore the role of automation in the new system. In other words, they began to explore the boundary between the work that should be done by a computer and the work that should be done by a human. Choosing the proper system boundary and the proper mix of human work and computer work is the purpose of the Evaluation of Alternatives Phase.

The choice of a suitable boundary and work mix is based upon the value of each alternative. Determination of value involves an analysis of system benefits, costs, and risks [4]. Analysis of system benefits can be difficult because some benefits are not easily quantified and some benefits are not certain to occur.

In considering the automation of Job Cost Reports, Sam identified the elimination of a full-time report preparer whose salary was $18,000. This benefit is easily quantified and easily obtained.

Sam also indicated that his staff of engineers and technicians would save one hour per day from the elimination of the manual reports. This benefit is also easily quantified, but it is not certain to occur. It is important that the analyst assess the likelihood that a benefit will be obtained and the form in which it will be obtained in addition to the dollar amount of the benefit. Pete's question about what would be done with the time the engineers save on Job Cost Reports is crucial.

Finally, Sam and Pete discussed some benefits that are not easy to quantify. The cost of introducing a new person to a project is real, but it is difficult to estimate this cost accurately. Sam talked about understanding the operation of his department better to make better decisions. How much are better decisions worth? No one can say for sure.

Benefits such as better decisions are called **intangible benefits** because they are difficult to quantify without experience. If a benefit is easy to quantify, then it is called a **tangible benefit**. In this part of the Evaluation of Alternatives Phase the analyst tries to identify both tangible and intangible benefits, to determine the dollar value of the benefits wherever possible, and to assess the likelihood the benefits will occur and the form they will take. Costs are considered later in the life cycle when the physical model of the new system is developed.

In Chapter 4 the decision to continue past the Problem and/or Opportunity Analysis Phase was based upon the promise of improving performance quality and efficiency and cost control mechanisms through an information system. These information system functions usually provide tangible benefits.

Naturally, the more tangible benefits there are, the easier it is to assign a dollar value to the new system and the easier it is to commit resources to the development effort. When the benefits of the proposed system are mostly intangible, then the development team adopts an incremental approach as Pete suggested for the vendor reorganization requirement. Remember, intangible does not mean nonexistent; it means difficult to quantify without experience.

Before you begin the next section, read again the previous section on turnover. Remember that the analyst tries to identify the benefits, the value of the benefits, the likelihood the benefits will occur, and the form the benefits will take. Try to identify the pattern and the motives for Pete's questions. In the next section try to anticipate what Pete's questions will be.

The Value of Automation—
Monthly Budgets

Pete began the meeting with a recap of his previous session with Sam. "We have discussed items 2 and 5 from the Requirements Model, Figure 6.2, and we have seen that the automation of Job Cost Reports brings substantial benefits to the company in terms of cost savings and in terms of upgrading the activities of professional employees. What about item 3 in the Requirements Model, Monthly Budgets?" asked Pete.

"Let me put it simply," said Sam. "In this department we spend one million dollars per year on payroll, another 1.1 million on subcontractors, and a half million on equipment and materials. Every year we exceed budgeted expenditures by approximately 4%. That's $104,000, most of which we swallow. By the time we find out about excess expenditures, it is usually too late to go to either the vendor or the customer for help. Most of the time we can't pinpoint the exact cause of the overrun anyway.

"A lot of the experienced people accept the overruns as a fact of life, but I think technology can improve our situation. I want to be able to plan the activities for the month before they happen instead of reacting to them after the fact."

"What's the likelihood that you'll save anything with the Monthly Budgets?" asked Pete.

"I can't see why we shouldn't save it all eventually," said Sam. "I am already requiring vendors and subcontractors to submit monthly schedules with their bids, and I have caught a few things. At the moment I am working on tightening up payroll expenses.

"The difficulty is the same as I face with the Job Cost Reports. I use the same data over and over again but never in the same way twice. At any given moment I usually need access to only a few pieces of information. The problem is finding those few pieces and organizing them in the right way. Right now I spend more time sifting through the information than using it to run the department."

The Value of Automation—
Wrapping It Up

"What about the value of Vendor Control, which is item 4 in the Requirements Model?" asked Pete.

"You tell me," said Sam. "We spend 1.6 million dollars per year on equipment, materials, and subcontractors. A percentage point decrease in that amount equals $16,000. Is it worth pursuing?"

"I do not know yet," said Pete. "These discussions are establishing the potential worth of the new information system. The potential worth determines the alternatives I will present to you. I will present the strengths and weaknesses of each alternative as well as the costs, and you will decide what is worth pursuing and what is not.

"The only requirement we have not yet discussed is number 1," said Pete. "What work on the Analysis DFD, Figure 5.23, is more suited to a computer than a human?"

"Let's see," said Sam. "Completing the payroll roster and payment authorization vouchers is fairly simple. I compile 22 time logs into a roster every two weeks. That takes about an hour. I process about 100 payment authorizations every month. That takes a couple of hours each week, but I don't mind doing it. It keeps me on top of what's going on.

"Betsy Klein, our administrative assistant, does the postings to the job cost ledger. Getting the data in for the first time is not difficult. It's getting the information out that causes the problems, and I am satisfied with what we have discussed about that.

"What about process 3.   MONITOR ANNUAL BUDGET on the Analysis DFD?" asked Sam. "Will we keep it or throw it away?"

"I won't know that until I talk to Ed Henderson in Accounting," said Pete. "I have an appointment with him tomorrow. I'm hoping the accounting system will be able to provide some of the things we need."

"In processes 4 and 6 on the Analysis DFD, each situation is unique," said Sam. "Betsy does our requests for bids on a word processing system, which runs very well. We've already noted the need for Vendor Control in process 4. I don't see anything else."

"OK," said Pete. "Thanks for the time. I will be meeting with Ed tomorrow; then we'll put together some alternatives. If you are free, we can do that on Monday, and you should be able to report to your president by Wednesday."

"That's fine," said Sam.

"Before I go, give me some background on the personal computers I see around the department," said Pete.

"We have a network with five stations in the department," said Sam. "Betsy Klein is the most knowledgeable user. She does everything on her station, including her appointment calendar. We've all learned a lot from her. I have a station also. I use it for word processing and spreadsheets. The other three are used by the engineers for a variety of things."

"Thanks," said Pete.

Developing the Automated Portion of the New System

During his meeting with Ed, Pete learned that the minicomputer-based accounting system did not handle any job cost information, so the new system would have to address all of the items in the Requirements Model, Figure 6.2. Pete used the Requirements Model and his notes from his conversations with Sam to compile the chart that appears as Figure 6.6.

The Design DFD

Pete began with requirements 2 and 3 because of the large benefits and the high likelihood of obtaining them. Both Job Cost Reports and Monthly Budgets would require a transaction processing component to maintain the data stores for the reports. Pete liked to start the design process by sketching a data flow diagram for this transaction processing component. He called the diagram a Design DFD, and he followed a process similar to the one he used in developing the Analysis DFD [3].

He began by including the required outputs of the new system and the inputs of the current system in a Context Data Flow Diagram. As in the Analysis Phase, the data flows represent paths for information that crosses

Requirements Model Reference	Benefit/Yr	Likelihood
2. Job Cost Reports	18,000	Certain
	125,000	High
3. Monthly Budgets	104,000	High
5. Professional Development Plans	10,800 each	Even
4. Reorganize Vendors	16,000	Uncertain

FIGURE 6.6

Pete's summarization of the benefits of automation.

the system boundary. In this case the data flows represent paths for information going into and coming out of some kind of computer. The diagram appears as Figure 6.7.

Identifying the Central Data Stores

Pete grouped the job cost data flows into two processes connected by a data store which he called EXPENSES. He assumed the contents of EXPENSES would be similar to the contents of the job cost ledger in the current system and noted this on the diagram. His diagram appears as Figure 6.8.

Pete added the Monthly Budget data flows to his diagram. The Budget vs. Actual Report needed the EXPENSES data store, but it also needed budget information. Pete decided to add a second data store, which he called BUDGETS. The updated diagram appears as Figure 6.9.

Pete returned to Figure 6.6 to consider adding other requirements to his design. Professional Development Plans appeared next so he brought out his DFD for process 7. MAINTAIN PROFESSIONAL DEVELOPMENT PLANS, Figure 6.3.

Pete added the Engineer's Activity Summary output data flow to both the Context and the Design DFDs. Since Figure 6.3 showed the time logs as the source of the Engineer's Activity Summary, no additional input data flows were required. Pete's notion of the EXPENSES data store, however, expanded to include a combination of the job cost ledger and the time logs from the current manual system.

Pete agreed with Sam's assessment that the Activity Summary data flow and the engineer's activity data store needed to be automated and that everything else in this process should be manual, but he included both data

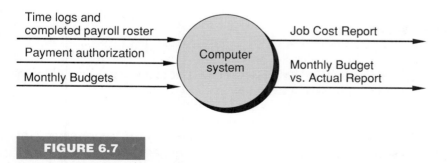

Time logs and
completed payroll roster

Payment authorization

Monthly Budgets

Computer system

Job Cost Report

Monthly Budget
vs. Actual Report

FIGURE 6.7

First draft of Automated System context DFD.

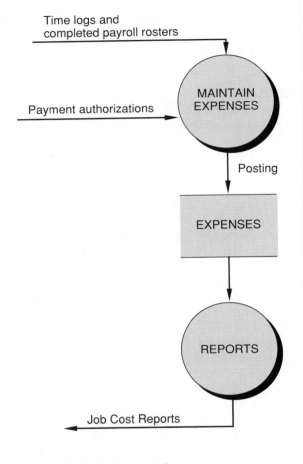

Time logs and
completed payroll rosters

Payment authorizations

MAINTAIN
EXPENSES

Posting

EXPENSES

REPORTS

Job Cost Reports

EXPENSES = Job cost ledger

FIGURE 6.8

The job cost portion of the new system.

stores from Figure 6.3 in the Design DFD. Decisions regarding physical implementation of the design would be made at a later date.

The current version of Pete's design appears in Figure 6.10. He saw the familiar pattern of a transaction processing and/or reporting system emerging. The BUDGETS and PLANS data stores represented the master; they contained background information on jobs, general ledger accounts, and engineers. The EXPENSES data store represented day-to-day business transactions that would be reviewed against the background of the

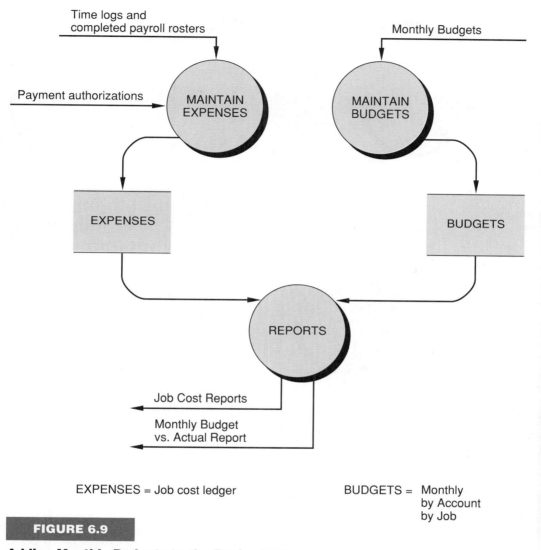

EXPENSES = Job cost ledger

BUDGETS =   Monthly
by Account
by Job

FIGURE 6.9

Adding Monthly Budgets to the Design DFD.

BUDGETS and PLANS data stores in the REPORTS process. Pete was not surprised at the emergence of the **master/transaction structure**; he had seen automated transaction processing and/or reporting systems replace manual systems in this way many times before.

Pete returned to Figure 6.6. Requirement 4. Reorganize Vendors appeared next. He reviewed the DFD for process 9. MAINTAIN VENDOR CONTROL DATA STORE, Figure 6.5. He could see no connection between

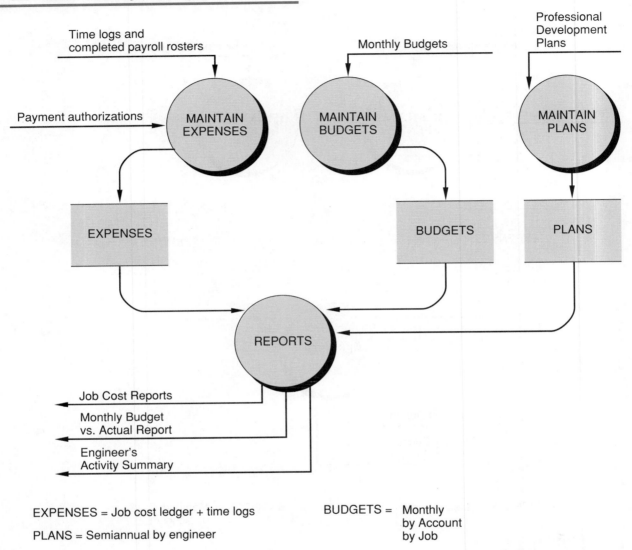

EXPENSES = Job cost ledger + time logs

PLANS = Semiannual by engineer

BUDGETS = Monthly
by Account
by Job

FIGURE 6.10

**Adding Professional Development Plans to
the Design DFD.**

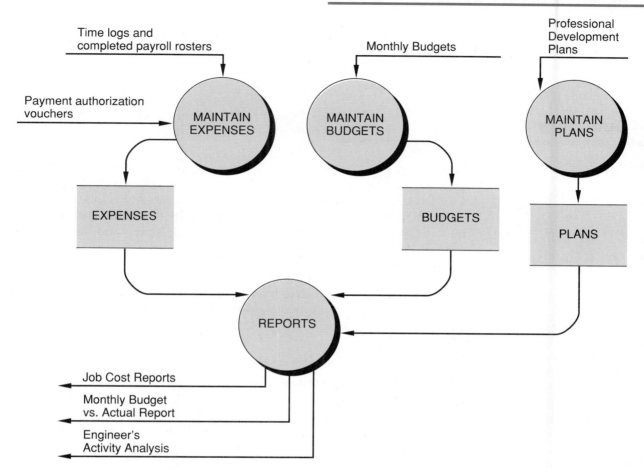

Time logs and
completed payroll rosters

Monthly Budgets

Professional
Development
Plans

Payment authorization
vouchers

MAINTAIN
EXPENSES

MAINTAIN
BUDGETS

MAINTAIN
PLANS

EXPENSES

BUDGETS

PLANS

REPORTS

Job Cost Reports

Monthly Budget
vs. Actual Report

Engineer's
Activity Analysis

EXPENSES = Job cost ledger + time logs

PLANS = Semiannual by engineer

BUDGETS = Monthly by Job by Account

Note: Vendor Control?

FIGURE 6.11

The Final Design DFD.

this process and his design data flow diagram, so he wrote the words "Vendor Control?" in the lower right-hand corner of Figure 6.10. The changes to Figure 6.10 are shown in Figure 6.11.

Summary

In this section Pete drew a data flow diagram for the automated portion of the work of the new system. As a logical model the Design DFD represents *what* work should be done and not *how* the work should be done. The main purpose of drawing the Design DFD is to identify the central data stores of the new system, the relationships among the data stores, and the potential sources of the data stores' contents.

In the next chapter Sam and Pete will review the Design DFD to determine various ways in which the specified work can be done. Pete was confident that an automated system configuration could be developed within the bounds of the benefits uncovered in his work with Sam.

THOUGHT QUESTIONS
In Your Opinion. . .

1. What project can you think of for which the physical model of the current system must be developed?

2. What project can you think of for which the logical model of the current system need not be developed?

3. Do you think it is possible to separate the physical from the logical completely? Do you think it is necessary? Do you think it is desirable?

4. What are the similarities and differences of the sessions between Sam and Pete presented in this chapter and those presented in Chapter 5? What do you think is the most difficult problem facing the systems analyst at this stage of the life cycle? What new communication techniques is Pete employing? Do you think they are suitable for the work of the sessions? Why or why not?

5. Do you think a computer will ever replace Joe Sindora's expertise regarding vendors? What do you see the role of computers to be in situations such as Sam's vendor base problem?

6. Consider a systems development project that requires 24 person-months to complete. In light of Sam's comments on the cost of turnover,

what is the best assignment of workers? One person for 24 months? 24 people for one month? Something in between? What definition of "best assignment of workers" did you use?

7. Think about systems that you see every day. Name one that you think provides tangible benefits that were certain to occur. Name one that you think provides tangible benefits that were not certain to occur. Name one that you think provides primarily intangible benefits.

References

1. Bostrom, R. "Successful Application of Communication Techniques to Improve the Systems Development Process," *Indiana University Working Paper*, 1987.

2. DeMarco, T. *Structured Analysis and System Specification.* New York: Yourdon Press, 1978.

3. Eckols, S. *How to Design and Develop Business Systems.* Fresno, CA: Mike Murach and Associates, Inc., 1983.

4. Keen, P. "Value Analysis: Justifying Decision Support Systems," *MIS Quarterly*, Volume 5, Number 1, March 1981, pp. 1–15.

HORATIO&CO.

Moving Toward the Design DFD

Experience is another important characteristic of the successful systems analyst. In this assignment section you will get a chance to apply the tools and techniques of this chapter to the Horatio & Co. Credit Union System scenario. In this scenario a manual system is being replaced, but cost savings is not necessarily the dominant criterion. You will notice other differences as well as similarities between the two scenarios. Be creative. The credit union system scenario is designed to use and expand your base of experience regarding the automation of manual transaction processing and reporting systems.

Exercises

1. Review your answers to exercise 7 of Chapter 4 and exercise 6 of Chapter 5. How did the similarities and differences you noted in those exercises manifest themselves in the Requirements Model for the credit union system?
2. Determine the boundary of the new credit union system from the Requirements Model.
3. Determine the work of the new system that is best suited to automation through an analysis of the benefits of automation, the likelihood these benefits will occur, and the form these benefits will take. Be sure to include both tangible and intangible benefits.
    Assume that Camille Abelardo earns $25,000 per year and that all accounting clerks who do credit union work earn $8.00 per hour.
4. Organize your work in exercise 3 into a chart resembling Figure 6.6.
5. Based upon your work in exercises 3 and 4, develop a Context DFD for the automated portion of the new system. See Figure 6.7.

6. Complete a Design DFD to identify the central data stores and the transaction processing structure of the automated portion of the new system.

CHAPTER 7

Evaluating Solution Alternatives

Objectives

In this section Pete helps Sam to decide how the work specified in the Design DFD will be done. The process ends with the development of a physical model of the new system, Figure 6.1.

At some point in this chapter the cost control project shifts from analysis work to design work. The boundary between the Analysis Phases and the Design Phases of the systems development life cycle is not easy to identify. Like logical and physical modeling, analysis and design very often take place simultaneously.

The decisions regarding the physical form of the new system involve striking a balance between system features, user responsibility, and cost. The Designer's Tradeoff Chart presented in this chapter supports that decision. Through your experiences with the Designer's Tradeoff Chart, you will come to realize why some authors believe that **tradeoff** is the most important word in the systems analyst's vocabulary [10].

The specific objectives of this chapter are

1. To generate alternative models of the physical implementation of the Design Data Flow Diagram;

2. To arrive at a feasible physical form for the automated portion of the new system through tradeoffs involving system features, user responsibilities, and cost.

Generating the Alternatives

Pete arrived at Horatio & Co. early on Monday morning. He discussed his summarization of the benefits of automation, Figure 6.6, and the Design DFD, Figure 6.11, with Sam.

Sam agreed that Job Cost Reports, Monthly Budgets, and Professional Development Plans exhibited a strong logical connection. He was disappointed that the reorganization of the vendor base did not share this connection. He questioned Pete about the fate of the vendor base project.

"Based upon the logical model of the work specified in the new system requirements, the vendor base project is a separate entity," said Pete. "We can pursue it at the same time as the cost control project, or we can wait until we achieve some definite results on the cost control system."

"Let's wait," said Sam, "but let's be careful not to forget about it completely."

"Fine," said Pete. "Today, then, we will begin to think about the physical implementation of the Job Cost Reports, the Monthly Budgets, and the Professional Development Plans. The **Designer's Tradeoff Chart** is something I have developed over the years to help with this part of the project.

"Computer-based information systems are made up of five components: hardware, software, data, procedures, and personnel. The Designer's Tradeoff Chart presents a detailed listing of options for these five components. We'll use it to determine a feasible configuration for the new system."

Pete continued, "We always consider feasibility on three levels. **Operational feasibility** refers to how well a specific configuration matches the people in the user/management group. On this project we are working with a group of enthusiastic technical professionals. If they are willing, they could accept a high level of responsibility in return for a sophisticated and flexible system. Such a system might not be feasible in another organization.

"**Technical feasibility** is the second level. This refers to the ability of the technical resources to support a given configuration.

"Finally, **economic feasibility** refers to the traditional cost-benefit analysis [4]. Will the tangible and intangible system benefits be worth the cost?"

The Designer's Tradeoff Chart for the Current System

Pete asked Sam to look at Figure 7.1. "This is how your current system looks on the chart. We'll start with the ACCESS section. For job cost information you go through an intermediary, Kim. For budget information, you use your files, which is one shared direct access point. Kim can prepare any report anyone needs, which is a big plus. In computer jargon that's known as ad hoc reporting. The minus is the waiting time. Your job cost ledger is current to within a day; your Budget vs. Actual Report gets to be a month old before it is replaced. You have no hardware or software responsibilities. You are responsible for your source documents and for data entry into the job cost ledger, and you must keep confidential information such as payroll data secure. Because of Kim and the manual files, no one needed to learn much to use the current information system."

"I used to think that our system was all bad," said Sam. "Now I see that there are some features that I do not want to give up."

"That's why we're here," said Pete." "Use the tradeoff chart to describe your ideal system by putting an X next to all the features you want."

Sam's Ideal System

"Well," said Sam. "The most important thing is to improve the TURN-AROUND on job cost. I want my own screen and printer, but I know there are some who do not want these things on their desks. So let's go with a combination of ACCESS options. Let's eliminate the month-old data under TIMELINESS, and leave everything else the same." The revised tradeoff chart appears in Figure 7.2.

Technical Feasibility

Pete looked at the chart. "Let's consider the technical feasibility of your ideal system first. If you do not want any hardware responsibility, then you can either use the central accounting system or go to an outside service bureau. Neither option is perfect.

"In talking to Ed, I learned that the accounting system can handle monthly budgets by general ledger account," continued Pete. "The bad

DESIGNER'S TRADEOFF CHART

ACCESS

	Direct access points on desks
	Shared direct access points near desks
NOW	One shared direct access point
NOW	Access through intermediary

FLEXIBILITY OF OUTPUT

NOW	Ad hoc reporting (selection of presentation and data)
	Several predetermined presentations with selection of data
	One predetermined presentation with selection of data
	Several predetermined presentations with no selection of data
	One predetermined presentation with no selection of data

TURNAROUND TIME ON OUTPUT

	While you wait (seconds)
	While you wait (minutes)
	Come back later (hours)
NOW	Come back later (days)

TIMELINESS OF DATA

	Up to the minute
NOW	One day old
	One week old
NOW	One month old or more

SYSTEM RESPONSIBILITIES

	Responsible for hardware and software
	Responsible for hardware
	Responsible for software
NOW	Responsible for neither hardware nor software

DATA RESPONSIBILITIES

	Responsible for initial data conversion
NOW	Responsible for source documents
NOW	Responsible for data entry
	Responsible for updating and purging
	Responsible for backup and restore
	Responsible for match with other company systems
NOW	Responsible for security and confidentiality

LEARNING

	Everyone must learn everything
	Someone must learn everything; everyone must learn something
	No one must learn everything; everyone must learn something
	Someone must learn something; others need learn nothing
NOW	No one need learn anything significant

FIGURE 7.1

Evaluation of the current system.

DESIGNER'S TRADEOFF CHART

ACCESS
X   Direct access points on desks
X   Shared direct access points near desks
  NOW   One shared direct access point
  NOW   Access through intermediary

FLEXIBILITY OF OUTPUT
X NOW   Ad hoc reporting (selection of presentation and data)
               Several predetermined presentations with selection of data
               One predetermined presentation with selection of data
               Several predetermined presentations with no selection of data
               One predetermined presentation with no selection of data

TURNAROUND TIME ON OUTPUT
X   While you wait (seconds)
               While you wait (minutes)
               Come back later (hours)
  NOW   Come back later (days)

TIMELINESS OF DATA
               Up to the minute
X NOW   One day old
               One week old
  NOW   One month old or more

SYSTEM RESPONSIBILITIES
               Responsible for hardware and software
               Responsible for hardware
               Responsible for software
X NOW   Responsible for neither hardware nor software

DATA RESPONSIBILITIES
               Responsible for initial data conversion
X NOW   Responsible for source documents
X NOW   Responsible for data entry
               Responsible for updating and purging
               Responsible for backup and restore
               Responsible for match with other company systems
X NOW   Responsible for security and confidentiality

LEARNING
               Everyone must learn everything
               Someone must learn everything; everyone must learn something
               No one must learn everything; everyone must learn something
               Someone must learn something; others need learn nothing
X NOW   No one need learn anything significant

FIGURE 7.2

Sam's ideal system

news is that the accounting system has no provision for handling job cost information or ad hoc inquiries. And there are no job cost packages for sale that run on your minicomputer. Running the new system on the minicomputer means programming from scratch in COBOL."

"Is that bad?" asked Sam.

"It should be our last resort," said Pete.

"What do we do?" asked Sam.

"In this project, we are raising serious questions regarding the future of computing at Horatio & Co.," said Pete. "We do not want to set up your new system as an island. We need to stay in touch with the accounting system, but technology has advanced so far beyond the accounting system that it seems foolish to let it restrict us in our choices."

"Does that mean the service bureau?" asked Sam.

"I do not think so," said Pete. "At most, we are talking about a few hundred transactions per month for the new automated system, and at this time you have no need to communicate your data to anyone outside this building. Since you already have five personal computers in your department, I think we should look at a microcomputer-based system instead of a service bureau."

Trading Off

"What does that do to the tradeoff chart?" asked Sam.

"You have a lot more responsibility," answered Pete. "You will be responsible for running the hardware. Whatever software you use, you will be the only group in the company using it. You will also be responsible for maintaining your own database, including keeping it in synch with the accounting system."

"I don't like that at all," said Sam. "If there is a difference between the two systems, we'll never be able to find the cause."

"I know," said Pete. "It's a real problem. We can try running your new system independently for awhile. Once we are sure of the data we want to capture, we can ask Ed to develop some programs for the minicomputer that would allow Betsy to enter the data needed by both systems. Each day you would transfer newly entered transactions from the minicomputer to a microcomputer here over a connecting line."

"Can that be done?" asked Sam.

"Yes," said Pete. "With the proper hardware and software, a microcomputer can emulate a minicomputer terminal and accept minicomputer data files. The volume of data in this case is within the bounds of this type of file transfer. Maintaining 100% accuracy will require some programming, but it can be done.

"You have a decision to make, Sam," said Pete. "The only way to get the benefits you want is to go beyond the accounting system. The price you pay is increased responsibility and learning. For an information system that replaces your manual job cost ledger and provides monthly budgeting by job and general ledger account through a microcomputer-based system with ad hoc reporting capability, you face a tradeoff chart that looks something like this. In addition, you will begin the collection of the data for the vendor control data store." Pete wrote his description of the new computer system on the back of the chart. He modified Sam's ideal system based upon the discussion above. The modified tradeoff chart appears as Figure 7.3.

Operational Feasibility

As Sam reviewed the chart, Pete said, "In order to judge operational feasibility, we must determine if your department will use these features and if you can accept these responsibilities."

"Under LEARNING, you say someone must know everything," answered Sam. "Does that mean we will have to hire a computer expert?"

"I have been talking regularly to Betsy Klein, and I think she is capable of managing the system," said Pete. "If she is chosen, you will need to review her responsibilities to make room for the new assignment. Do you think she'd be interested?"

"I know she would," said Sam. "She is very anxious to see the results of our work. She told me the more computer responsibility she gets, the better."

"Once you settle upon a mix of features and responsibilities that is right for your organization, we will estimate the cost of implementation. If the cost is within your means based upon the benefits we uncovered earlier, then we are done. If the cost is too high, we go back to the tradeoff chart to either cut features or increase responsibilities in order to cut costs.

"If the cost of an acceptable features/responsibilities mix is low, then we can go back to the chart to add features or cut responsibilities, which would result in a better product, or the company can pocket the savings. We will continue to trade off features, responsibilities, and cost until we come up with the right combination."

"Under LEARNING, what is the something that everyone needs to learn?" asked Sam.

Pete answered, "We need to purchase a software package for the ad hoc reporting. Some packages allow the user to formulate a report request in

DESIGNER'S TRADEOFF CHART

<u>ACCESS</u>
X   Direct access points on desks
X   Shared direct access points near desks
  NOW   One shared direct access point
  NOW   Access through intermediary

<u>FLEXIBILITY OF OUTPUT</u>
X NOW   Ad hoc reporting (selection of presentation and data)
            Several predetermined presentations with selection of data
            One predetermined presentation with selection of data
            Several predetermined presentations with no selection of data
            One predetermined presentation with no selection of data

<u>TURNAROUND TIME ON OUTPUT</u>
X   While you wait (seconds)
            While you wait (minutes)
            Come back later (hours)
  NOW   Come back later (days)

<u>TIMELINESS OF DATA</u>
            Up to the minute
X NOW   One day old
            One week old
  NOW   One month old or more

<u>SYSTEM RESPONSIBILITIES</u>
X   Responsible for hardware and software
            Responsible for hardware
            Responsible for software
  NOW   Responsible for neither hardware nor software

<u>DATA RESPONSIBILITIES</u>
X   Responsible for initial data conversion
X NOW   Responsible for source documents
X NOW   Responsible for data entry
X   Responsible for updating and purging
X   Responsible for backup and restore
X   Responsible for match with other company systems
X NOW   Responsible for security and confidentiality

<u>LEARNING</u>
            Everyone must learn everything
X   Someone must learn everything; everyone must learn something
            No one must learn everything; everyone must learn something
            Someone must learn something; others need learn nothing
  NOW   No one need learn anything significant

FIGURE 7.3

Pete's modification to Sam's ideal system.

English-like sentences; others use a more rigid syntax. Whoever wants to generate an ad hoc report will have to learn how to use the reporting package."

"I have seen those packages. They're great, but not everyone here is going to take to formulating those commands. Can't we also have a menu system that allows somebody to pick reports by number?"

"Yes, you can," answered Pete. "I'll add predetermined presentations with selection of data to the features. Everyone will still need to learn something, but that something can be as little as how to use a menu-driven system."

"Will that raise the cost?" asked Sam.

"Yes," said Pete. "You'll make the choice when we decide on the application software." The new version of the tradeoff chart is shown in Figure 7.4.

Sam looked at the tradeoff chart. "I do not like the responsibility for matching our data with the accounting system, but as long as you say we can develop a common entry and control point, I'll live with the responsibility for awhile. I am anxious to see the cost of this alternative."

Beginning Economic Feasibility Analysis

"We will estimate the cost of this alternative by reading down the tradeoff chart," said Pete. "Under ACCESS, you indicate the need for several direct access points. There is no reason why we cannot use your existing network. I talked to the engineers. They said that two of their three machines are in use often, but rarely are all three used. We can add another station for $2,000; more stations will require additional networking hardware."

"Let's go with one additional station for now," said Sam.

"The FLEXIBILITY feature on the tradeoff chart concerns the application software," said Pete. "I have already said that programming our own application software should be a last resort. Job cost accounting has become a standard computer application; there are many microcomputer-based packages available. Let's consider this option first."

The Question of Software Acquisition/Development

"I know those packages," said Sam. "We looked at several of them, and bought one a few months ago. It's a good accounting package, but I need

DESIGNER'S TRADEOFF CHART

ACCESS
X	Direct access points on desks
X	Shared direct access points near desks
NOW	One shared direct access point
NOW	Access through intermediary

FLEXIBILITY OF OUTPUT
X NOW	Ad hoc reporting (selection of presentation and data)
X	Several predetermined presentations with selection of data
	One predetermined presentation with selection of data
	Several predetermined presentations with no selection of data
	One predetermined presentation with no selection of data

TURNAROUND TIME ON OUTPUT
X	While you wait (seconds)
	While you wait (minutes)
	Come back later (hours)
X NOW	Come back later (days)

TIMELINESS OF DATA
	Up to the minute
X NOW	One day old
	One week old
NOW	One month old or more

SYSTEM RESPONSIBILITIES
X	Responsible for hardware and software
	Responsible for hardware
	Responsible for software
NOW	Responsible for neither hardware nor software

DATA RESPONSIBILITIES
X	Responsible for initial data conversion
X NOW	Responsible for source documents
X NOW	Responsible for data entry
X	Responsible for updating and purging
X	Responsible for backup and restore
X	Responsible for match with other company systems
X NOW	Responsible for security and confidentiality

LEARNING
	Everyone must learn everything
X	Someone must learn everything; everyone must learn something
	No one must learn everything; everyone must learn something
	Someone must learn something; others need learn nothing
NOW	No one need learn anything significant

FIGURE 7.4

**The tradeoff chart after the addition of
report menus.**

more flexibility in the reporting. In order to see engineers' activity, I had to set up each engineer as a vendor, which meant assigning and remembering identification numbers. It was more trouble than it was worth. Our use of the job cost data is not the same as that of the accounting department."

"I agree," said Pete. "The next logical step is to consider using the accounting package to maintain the data and running ad hoc reporting software on the package's database."

"Can you do that?" asked Sam.

"It depends upon the package," said Pete. "In this case we're going to use Horatio's accounting system for data entry and maintenance eventually, so there is no need to consider another package. If this system catches on with other departments, then the company needs to look into replacing the current accounting system with something that will allow easier access for MIS and DSS applications. As I said, we are raising serious questions about the future of computing at Horatio & Co. This project is really a pilot study in that much bigger effort."

"So, what do we do?" asked Sam.

"It looks like we build the software," said Pete. "Betsy Klein knows dBASE III PLUS very well. Like most sophisticated dBASE users, she has taken advantage of the many add-on products that are available for the package. Betsy uses a computer-aided software engineering (CASE) tool for dBASE called GENIFER. The combination is very powerful. Betsy has expressed an interest in participating in this project. I think she can build your system with dBASE III PLUS, GENIFER, and a little help from me in the early stages."

"Does this option satisfy all three feasibility requirements?" asked Sam.

"Yes," said Pete. "dBASE III PLUS runs on your network and can easily process a few hundred transactions per month. GENIFER needs to run on Betsy's station only, so the idea is technically feasible."

"Will dBASE III PLUS handle the data transferred from Ed's system?" asked Sam.

"That's a good question," said Pete. "It will."

"What about cost?" asked Sam.

"A network version of dBASE III PLUS lists for about $1,500," said Pete. "You already own a copy of GENIFER."

"And operational feasibility?" asked Sam.

"Betsy is willing and able to build the system," said Pete. "That is critical. We are lucky to have a competent system builder on staff. dBASE III PLUS has a command language and a report writer for ad hoc inquiries. It's not the most powerful package on the market, but it is simple to learn."

"Is dBASE III PLUS an example of a fourth-generation language?" asked Sam.

"Yes, a simple one," answered Pete. "A fourth-generation language (4GL) is a combination database-management/application-development tool. At an elementary level, dBASE III PLUS satisfies that definition."

"And GENIFER is called a CASE tool?" asked Sam.

"Yes," said Pete. "CASE tools maintain a database of information about a project and automate many systems development processes. GENIFER supports the later phases of the life cycle, particularly the Buy/Build Software Phase. Other CASE tools are available for the earlier phases [5].

"If you are willing to make a commitment of your time, the feasibility of a 4GL/CASE environment for this project allows us to develop our software through prototyping. Prototyping involves developing a working system very quickly, which you and your staff use and evaluate. Your feedback goes into enhancing the prototype, and we continue to refine the system until we have an operational product."

"How much time is required?" asked Sam.

"For estimating, I use five iterations of the prototyping process, which require two days each from the user and the builder over a two-month period. You'll decide at the end of each step if the added benefits are worth the cost," said Pete.

"Ten days out of the next 40 work days is 25% of our time," said Sam. "We'll have to plan. What happens if we cannot make the commitment?"

"We would still work within the 4GL/CASE environment, but not as effectively," said Pete. "Prototyping evolved as a development method because it exploits the 4GL/CASE technology to provide a working system as quickly as possible" [7].

"What if the 4GL/CASE environment were not feasible?" asked Sam.

"In a less sophisticated programming environment, one cannot afford to make many changes," answered Pete. "If we could not use a 4GL or a CASE tool like GENIFER, then we would develop a detailed design of the software before doing any programming, and the system would be delivered to you when all of the programming was complete."

"That does not sound as flexible as the prototyping approach," said Sam.

"It's not," said Pete. "But the cost of programming in a third-generation language such as COBOL is so high that the developers really have no choice."

"When is the 4GL/CASE environment not feasible?" asked Sam.

"Look at the three criteria," said Pete. "Microcomputer-based 4GLs are inexpensive; mainframe versions are not. If the requirements of the system dictate mainframe use, a 4GL may not be economically feasible. Since 4GLs are relatively new, it's possible to find an organization in which no one knows how to use a 4GL effectively. As I said, we're lucky to have Betsy for this project. Finally, 4GL programs do not yet run as efficiently as COBOL programs, although they are getting better all the time. If a system processes huge volumes of data, the 4GL may not be able to provide the necessary processing power."

"I think we can manage to commit the time," said Sam. "Let's see if we can estimate the cost of the commitment. My time plus Betsy's time is worth $400 per day, times 10 days is $4,000. It may also cost us some overtime and temporary help fees."

"So," asked Pete, "can we assume we are using prototyping to build the software component of the cost control system?"

"Yes," said Sam. "It will not be easy to find the time, but let's assume prototyping for the moment."

Continuing Economic Feasibility Analysis

Pete continued, "The next item on the tradeoff chart is TURNAROUND. You will be able to get the turnaround you need from the current file server on the network. If Betsy continues to do data entry each day, the TIMELINESS should not be a problem. Her time for data entry with the computer system and the current job cost ledger should be about the same.

"The next items are SYSTEM and DATA RESPONSIBILITIES. You'll have to plan on about 10% of Betsy's time devoted to managing the operation of the system on an ongoing basis. This includes controlling the match with the accounting system. There will be more security and confidentiality concerns than with the manual system. dBASE III PLUS provides standard network database protection features that Betsy can use to encrypt sensitive data and restrict access to it."

"Betsy costs us $30,000 per year, including fringes," said Sam. "Is that competitive with the market or will I have to worry about losing her too?"

"The salary is competitive, but you still have to worry about losing her," said Pete. "You or someone else will have to learn as much as you can to provide some backup for her."

"So, $3,000 of Betsy's total compensation must be charged to this system each year," said Sam. "In addition, we have to consider the cost of initial data conversion, programming the minicomputer for the new data entry procedures, and integrating the new automated system with our manual procedures." "Ed Henderson estimates $2,500 for the data entry programming. Integrating the automated system with the manual procedures should take one day with you and Betsy working on it," said Pete.

"That's $400," said Sam.

"**Inital data conversion** involves some choices," said Pete. "Look at the Design DFD, Figure 6.11. Let's assume that the EXPENSES data store is equivalent to the manual job cost ledger. At present, 10 engineering jobs are underway. Each job generates 30 job cost ledger entries per month, and the average age of these jobs is about 12 months. 30 x 12, or 360, job cost ledger entries per job times 10 jobs means 3600 job cost ledger lines must be entered into the new system. I have a bid from a data preparation service to key the job cost ledger lines with verification onto diskette for $0.25 each, or $900."

"What are our options?" asked Sam.

"We could do the data entry here," said Pete.

"I do not have the available trained personnel for that," said Sam. "Does the preparation service provide accurate work?"

"We are lucky there," said Pete. "The job cost ledger is in good shape. With the verification of the entries, we will get accurate data."

"Let's use them," said Sam. "What else?"

"We could convert only a few of the current jobs," said Pete.

"If anything, I would like to build a bigger file by entering some recently completed jobs," said Sam. "Will the prototyping process give us some insight on this question?"

"Yes," said Pete.

"Then let's enter the 10 current jobs and wait to use the prototype to decide if we need to convert records of completed jobs as well," said Sam.

"That sounds reasonable," said Pete.

"Once the job cost ledger lines are entered into the system, we will have to spend some time standarizing things like the source and description. Betsy Klein has collected a wealth of material on building a job cost data-base from trade publications."

"Shouldn't we standardize the data first?" asked Sam.

"No," said Pete. "We can use dBASE III PLUS to help with that. We can sort records on various keys, print proof lists, and make corrections much easier with computer records than with manual records. That's the advantage of working in a fourth-generation environment."

"Are there other data conversion considerations?" asked Sam.

"Yes," said Pete. "Look at the Design DFD, Figure 6.11. You must prepare Monthly Budgets by Job and general ledger Account and Professional Development Plans for each engineer."

"Most of that is done already," said Sam. "I think the budgets should be in a computer file and the plans should be kept in manual files. It should take me 2 days to complete the preparation. That's worth $600."

"That's a good way to start," said Pete. "We need the budgets in a computer file for the Budget vs. Actual Reports, but we do not know enough about the planning system to determine what should be stored and how. The prototyping process will tell us more."

"What about the last item on the tradeoff chart, LEARNING?" asked Sam.

"Well," said Pete, "everyone is going to need about an hour of training to learn the menu-driven system; Betsy can do the session. Figure $500 for that. People like you who want to learn ad hoc reporting will need to spend about 20 hours doing so. Four-hour seminars with Betsy, once a week, for five weeks would be ideal."

"I'll have to work that into the assignments," said Sam. "I estimate that 10 people will be interested. That's 2.5 days at an average of $200 per day times 10 people. That's $5,000. What about your fee, Pete?"

"I'll have to do some training with Betsy on prototyping and help her through the first iterations of the prototyping process. Estimate $4,000 for me."

"Also, we cannot forget the cost of this analysis," said Sam. "Five days of my time and your fee comes to another $4,000."

Sam totaled the figures he had written on his notepad. The cost estimate was $24,650 for the first year plus $3000 per year, after the first year, for on-going management. See Figure 7.5.

"This is an easy decison," said Sam. "If we realize the $18,000 per year salary savings only, we'll be ahead by the second year. Any other benefits we achieve go right to the bottom line, and in addition, we're helping the company chart its computing course for the future. Could we ask for more?"

"Do you want to go back to the tradeoff chart to improve the system?" asked Pete.

"Why?" said Sam. "This is what I wanted. I can go to the president with specific means to achieve my plan and, more importantly, specific figures on the value of what I want to do."

Summary

The considerations of the Evaluation of Alternatives Phase of the systems development life cycle, Chapters 6 and 7, are summarized in Figure 7.6.

In deciding the new system boundary and the role of automation, the development team uses the Requirements Model as a guide. Each item in the Requirements Model is reviewed to determine the benefits of automation.

```
Additional network station                       $2,000
Minicomputer-to-PC line                          negligible
dBASE III PLUS network version                   $1,500
Software development (Sam and Betsy)             $4,000
System manager                                   $3,000 per year
Minicomputer data entry programming              $2,500
Minicomputer-to-PC communication software          $250
Integrate manual and automated procedures          $400
Conversion of job cost ledger entries              $900
Preparation of Monthly Budgets                     $600
Training (menu system)                             $500
Training (ad hoc reporting)                      $5,000
Software development (Pete)                       $4,000


TOTALS
New system startup                               $24,650
Ongoing                                          $3,000 per year

This analysis                                    $4,000
```

FIGURE 7.5

Tradeoff chart costs.

Once the benefits of automation are quantified for each item in the Requirements Model, the analyst develops the Design Data Flow Diagram to organize automated processes and data flows into a logical pattern, and to identify the central data stores and the transaction processing structure of the system. In this work the analyst designs any processes and data stores he or she sees fit to use to accomplish the specified work; the analyst is not bound by the physical reality of the current system in developing the Design DFD.

The analysis of software acquisition/development methods should examine prewritten software packages first. If a satisfactory package is not available, then the analyst considers developing the software. If a fourth generation/CASE environment is feasible and the user/management group is willing to commit the necessary time, prototyping should be used. If prototyping is not feasible, then the detailed design approach should be used.

The boundary of the new system, which is determined by the processes of the Analysis DFD included in the Requirements Model and the new processes specified in the Requirements Model (Chapter 6)

The role of automation in the new system, i.e., the mix of the work of the new system assigned to the computer and the work assigned to humans (Chapter 6)

The value of tangible and intangible benefits resulting from the implementation of the new computer system, the likelihood these benefits will occur, and the specific forms these benefits will take (Chapter 6)

The chief data stores of the new system, the relationships among the data stores, and the potential sources of the data stores' contents (Chapter 6)

The transaction processing structure of the new system (Chapter 6)

The software acquisition/development methodology, i.e., prewritten software packages, prototyping, or detailed design (Chapter 7)

The physical form of the computer system, i.e., the specific mix of features, responsibilities, and cost that best meets the user's needs, capabilities, and expectations (Chapter 7)

FIGURE 7.6

Considerations of the Evaluation of Alternatives Phase of the systems development life cycle.

Finally, the physical implementation of the new automated system is determined. The development team uses the Designer's Tradeoff Chart to trade off features, user responsibilities, and costs to arrive at an acceptable physical model of the new system. The tradeoff chart serves as a checklist of points to consider as well as a decision-making aid.

In Chapters 5 through 7, you saw the development team move through the Analysis DFD, the Requirements Model, the Design DFD, and the Designer's Tradeoff Chart. This pattern, Figure 6.1, is based upon an analysis and design theme developed by DeMarco [2]. It provides a useful guide to the automation of a manual transaction processing and reporting system.

Many organizations like Horatio & Co. begin their experience with computer-based information systems by automating transaction processing and reporting procedures. As you saw with the Job Cost Reports, computer processing offers a higher level of accuracy, better access, and faster response at less cost than manual processing. The tangible benefits are easy to identify, and often they pay for the automation project in a short time.

The automation of a manual current system is a good first experience in the education of a systems analyst. Today's systems analysts, however, face a broad spectrum of projects that range from the automation of manual current systems to the application of new technology to situations where no current system exists at all. From your experience with Sam and Pete, you should see that the systems analyst needs a broad range of business, technical, managerial, interpersonal, and creative skills.

THOUGHT QUESTIONS
In Your Opinion. . .

1. What do you think of the fact that Pete developed the Design DFD, Figure 6.11, without Sam? Is the analyst required to develop all models with the user? Why or why not?

2. Based upon the outcome in this case, what do you think is the value of developing the logical model of the new system?

3. Pete used the word *island* in one of his discussions with Sam. The term **islands of automation** has been used to describe disjointed systems developed independently throughout an organization. Do you think there is the potential to develop an island of automation in the cost control system project? Why or why not? Are islands of automation good or bad? What characteristics would you associate with an organization that has developed islands? Why do you think islands of automation occur? Who is responsible for making such decisions?

4. Sam and Pete decided to use a data preparation service for the inital data conversion. What factors do you think might preclude the use of such a service?

5. Betsy Klein has emerged as an important member of the cost control system development team. Redo Pete and Sam's tradeoff chart analysis assuming Betsy is not available for the project. Do you think Sam has any incentive to resist assigning Betsy to the project? Why or why not?

6. What do you think of Sam's method of evaluating the cost control system investment? It is called the payback period method. With it, the decision maker considers the length of time required to recoup the initial investment through the accrual of system benefits.

7. What other ways can you think of to evaluate the investment?

8. You used the initial prototype of the Horatio & Co. Cost Control System in Chapters 2 and 3. Does it seem like it would cost $25,000 to develop such a system even though very little hardware was purchased? Does it seem worth $25,000? How does this experience

compare with productivity software advertisements that claim end-users can develop their own applications quickly, inexpensively, and without the help of trained information systems professionals?

9. Think about ways to cut the cost of the system.

References

1. Brooks, F. *The Mythical Man-month.* Reading, MA: Addison-Wesley Publishing Company, 1975.

2. DeMarco, T. *Structured Analysis and System Specification.* New York: Yourdon Press, 1978.

3. Eckols, S. *How to Design and Develop Business Systems.* Fresno, CA: Mike Murach and Associates, Inc., 1983.

4. Emery, J. *Cost/Benefit Analysis of Information Systems.* The Society for Management Information Systems, 1971.

5. Gane, C. *Computer-Aided Software Engineering.* New York: Rapid System Development Inc., 1988.

6. Gremillion, L., and P. Pyburn. "Breaking the Systems Development Bottleneck," *Harvard Business Review*, March-April 1983, pp. 130–137.

7. Jenkins, A. M. "Prototyping: A Methodology for the Design and Development of Application Systems," *Spectrum*, Volume 2, Number 2, April 1985, pp. 1–8.

8. Jenkins, A. M. "Prototyping: A Methodology for the Design and Development of Application Systems—Part 2," *Spectrum*, Volume 2, Number 3, June 1985, pp. 1–4.

9. Keen, P. "Value Analysis: Justifying Decision Support Systems," *MIS Quarterly*, Volume 5, Number 1, March 1981, pp. 1–15.

10. Weinberg, G. *Rethinking Systems Analysis and Design.* Boston: Little, Brown and Company, 1982.

HORATIO&CO.

Evaluating Alternatives

The evaluation of alternatives for the credit union system is also based upon the Designer's Tradeoff Chart. Camille must consider access, flexibility of output, turnaround time on output, timeliness of data, system responsibilities, data responsibilities, and learning requirements.

For the software component, Camille decided upon a prewritten package running on a timesharing network. You will be asked to review this decision in the exercises.

Due to the increasing number of options, the software acquisition/development decision is not easy. Yesterday's solutions are no longer valid today, and today's solutions are certain to be called into question in the very near future.

Exercises

1. Complete a Designer's Tradeoff Chart for the manual operation of the credit union.

2. Using your own judgment, complete a Designer's Tradeoff Chart for an ideal system.

3. Using the work you did at the end of Chapters 2 and 3 with the credit union computer system, complete a Designer's Tradeoff Chart for the actual system Camille implemented.

4. Discuss differences and similarities between your ideal and the system Camille implemented.

5. The software component of the credit union system is a prewritten package running on a timesharing network. Read the article by Gremillion and Pyburn [6] cited in the chapter references, and discuss Camille's decision to use a software package instead of developing the software component of the system.

6. Discuss the procedures and decisions involved in the evaluation of a prewritten software package.

7. Prepare a list of tradeoff chart costs for the system Camille implemented. Use Figure 7.5 as a guide.

PART THREE

Systems Design

Part Three contains four chapters that cover the Design Phase and early Implementation Phase of the systems development life cycle. Design activities are based upon the five components of a computer-based information system: hardware, software, data, procedures, and personnel.

After discussing the organization of required resources into a project schedule and the accompanying presentation to management, Chapter 8 presents a method for the design of the hardware, procedures, and personnel components of a system. Chapter 9 discusses the design of the data and software components.

Chapter 10 explores the inner workings of the initial prototype of the Horatio & Co. Cost Control System, which is the system you used in Chapters 2 and 3. Finally, Chapter 11 discusses how the system user and the system builder work together to refine the initial prototype into a final operational version.

3

CHAPTER 8

Designing the New System

Objectives

In Chapter 7, Pete and Sam used the Designer's Tradeoff Chart to evaluate the economic, technical, and operational feasibility of information system alternatives. The final tradeoff chart, Figure 7.4, represents a rough design of the hardware, software, data, procedures, and personnel components of the automated portion of the system.

Sam Tilden is anxious to continue with the development of the cost control system. Before proceeding to the Design New System Phase of the systems development life cycle, however, Sam must report to the president of the company, Mr. Chapin.

If Mr. Chapin supports the continuation of the project, Pete and Sam will turn their attention to the details of the hardware, procedures, and personnel components of the automated portion of the new system. The Designer's Tradeoff Chart will serve as a useful guide in this work. Sam and Pete also will consider the integration of the automated system with the manual procedures of the Analysis DFD that were not chosen for automation.

Since prototyping is the software development methodology for the project, no detailed design work on the data and software components is necessary. If the design of the hardware, procedures, and personnel components does not uncover any difficulties, the project proceeds to the Buy/Build Software Phase of the systems development life cycle.

The specific objectives of this chapter are

1. To organize the resources required for the completion of the cost control system into a viable schedule;

2. To organize the results of the analysis phases of the life cycle into a report to senior management;

3. To observe the thoughts of senior management personnel regarding information system solutions to business problems and/or opportunities;

4. To design the hardware, procedures, and personnel components of a computer-based information system;

5. To determine the effect of the choice of software development methodology on the design of the data and software components of the new system.

Preparing for the Presentation to Management

To prepare for his presentation, Sam reviewed his previous report to the president, Figure 8.1. Sam remembered that Mr. Chapin was bothered by the lack of specifics in that report. The systems analysis he just completed with Pete provided specifics about his plan of action and the benefits and costs of that plan.

During systems analysis Sam discovered his uncertainty about his plan to reorganize the vendor base, so he moved the objectives and tactics regarding this item to the bottom of the list. He was also unhappy with the phrase "decreasing clerical costs" in the Business Objectives section. The phrase did not express the fact that the automated job cost reporting system would reduce the clerical activities of the engineers and technicians as well as replace clerical staff, so he expressed this fact in a new line in the Business Objectives section.

Sam included the list of yearly system benefits from Figure 6.6 in his report. He expressed the benefits in dollars and as a percentage of his department's budget on the new report. Recall from Chapter 6 that the engineering department spends one million dollars per year on payroll, another 1.1 million on subcontractors, and 0.5 million on equipment and materials, for a total of 2.6 million dollars per year.

Finally, Sam prepared a list of the costs of the proposed system. He broke the costs he identified using the Designer's Tradeoff Chart into five familiar components: hardware, software, data, procedures, and personnel. His new report to the president is shown in Figure 8.2.

Business Objectives

Maintain the current level of profit in the engineering department by

> decreasing clerical costs
> identifying cost overruns within 30 days
> decreasing cost of materials
> decreasing turnover

Business Tactics

1.   Eliminate manual preparation of job cost reports

2.   Institute a monthly budgeting system by job number and general ledger account to replace the current annual system by general ledger account only

3.   Reorganize the vendor base for higher volumes with fewer vendors

4.   Institute professional development plans for engineering personnel

System Objectives

Reports and inquiries on demand, including

> Budget vs. Actual Reports by job number and general ledger account on a monthly basis
> Job Cost Summary and Detail Reports
> Detail INQUIRY BY SOURCE of expense or vendor

Next Step

Proceed to the Model Building Phase of the systems development life cycle.

FIGURE 8.1

Sam Tilden's plan to improve business operations in the engineering department (repeated from Figure 4.2).

The Project Schedule

Sam used the tradeoff chart costs, Figure 8.2, to begin the development of the schedule for the remainder of the cost control system project. Project management was an important part of Sam's engineering training, and he was adept at scheduling and controlling project activities.

Business Objectives

Maintain the current level of profit in the engineering department by

> decreasing clerical costs
> reducing clerical activities of professional staff
> identifying cost overruns within 30 days
> decreasing turnover
> decreasing cost of materials

Business Tactics

1. Eliminate manual preparation of Job Cost Reports

2. Institute a monthly budgeting system by job number and general ledger account to replace the current annual system

3. Institute professional development plans for engineering personnel

4. Reorganize the vendor base for higher volumes with fewer vendors

System Objectives

Reports and inquiries on demand, including

> Budget vs. Actual Reports by job number and general ledger account on a monthly basis
> Job Cost Summary and Detail Reports
> Detail INQUIRY BY SOURCE of expense or vendor

System Benefits

REQUIREMENT	BENEFIT/YR	LIKELIHOOD
Job Cost Reports	$18,000 (0.7%)	certain
	$125,000 (4.8%)	high
Monthly Budgets	$104,000 (4.0%)	high
Professional Development Plans	$10,800 (0.4%) ea	even
Reorganize Vendors	$16,000 (0.6%)	uncertain

FIGURE 8.2

Sam's report to the president (page 1 of 2).

Costs (Start-up and ongoing)

HARDWARE

Additional network station	$2,000
Minicomputer-to-PC line	negligible

SOFTWARE

dBASE III PLUS network version	$1,500
Software development (Sam and Betsy)	$4,000
Software development (Pete)	$4,000
Minicomputer data entry programming	$2,500
Minicomputer to PC communication software	$250

DATA

Conversion of job cost ledger entries	$900
Preparation of Monthly Budgets	$600

PROCEDURES

Data entry replacing manual posting	even
Integrate automated and manual procedures	$400

PERSONNEL

System manager	$3,000 per year
Training (menu system)	$500
Training (ad hoc reporting)	$5,000

TOTALS

New system startup	$24,650
Ongoing	$3,000 per year
This analysis	$4,000

Next Step

Proceed to the Design New System Phase of the systems development life cycle.

FIGURE 8.2

Sam's report to the president (page 2 of 2).

A project schedule includes three components: the tasks to be performed, the resources required to perform the tasks, and the timing of the performance of the tasks. From Figure 8.2 and the notes of his meetings with Pete, Sam constructed the list of tasks and resources shown in Figure 8.3.

Because the events of the prototyping process are impossible to predict, Sam recognized the uncertainty of his schedule. Pete explained that the user controls the prototyping process and decides when the system is operational based upon his or her use of the current prototype [1]. Sam, however, did not feel comfortable about going to Mr. Chapin without, at least, an estimate of the time and resources required to complete the project. If problems arose or if additional revisions proved worthwhile, then modifications to the schedule could be made.

Pete estimated five versions of the cost control system would be developed. The first version would be evaluated by Sam alone, the second version by Sam and the engineers, and the third by Sam alone again. The accounting system data entry program and the accompanying file transfers would be included in the fourth prototype, which Betsy and Sam would test. Pete felt the revision of the fourth prototype should produce an operational system.

With the completion of the list of required resources, Sam turned his attention to the sequence of the events. The fourth prototype was critical. Outside data preparation and programming would have to be complete to develop the fourth prototype. In addition, the ad hoc reporting training would be beginning about the same time.

Sam organized his schedule into the **Gantt chart** [4] shown in Figure 8.4. He set the schedule based upon a maximum commitment of one-third of a person's time to the cost control system at any given moment. He did not feel that anyone, especially Betsy, could spare more time than that.

The labels on the Gantt chart show the time required to complete a given task. The lines show the time allotted for the task in consideration of the fact that no one is working on the cost control system full-time. For example, Betsy and Pete have ten days to complete task A: Prototype Training/Build Initial Prototype, a task that requires 2 days of commitment.

In setting the schedule, Sam tried to work up to the one-third commitment gradually by keeping commitments in the early stages of the project lighter than those in the later stages. The result was an estimate of eighty-three work days, or 16+ weeks, for the project, subject to the developments of the prototyping process.

Gaining Management Approval

Sam took a copy of his report to the president, Frank Chapin, the day before their scheduled meeting. In an accompanying letter he explained how he and Pete had arrived at the value of the benefits. He also explained the

A: Prototype training   Pete 2 days
     Build initial prototype   Betsy 2 days

B: Install network station
     & network dBASE III PLUS   Betsy 1 day

C: Test prototype   Sam 2 days each, versions 1, 2, 3
                                                          Engineers 2 days, version 2
                                                          Betsy 2 days, version 4

D: Revise prototype   Pete 2 days, version 1
                                                          Betsy 2 days each, versions 1, 2
E: Develop ad hoc
     reporting training   Pete 2 days

F: Integrate automated   Sam 1 day
     and manual procedures   Betsy 1 day

G: Develop Monthly Budgets   Sam 2 days

H: Standardize & proof   Pete 2 days
     job cost ledger entries   Betsy 2 days

I: Ad hoc reporting training   Engineers 2.5 days in 5 weeks
                                                          Betsy 3.5 days in 5 weeks (includes
                                                          preparation)

J: Key job cost ledger
     entries to diskette   Data prep service 1 day

K: Develop accounting system
     data entry program   Programmer 5 days

L: Test data entry program   Betsy 2 days
M: Build & test file transfers   Betsy 1 day

FIGURE 8.3

Cost control system resource requirements.

prototyping process and the role of this system as a pilot for a company-wide job cost system. Mr. Chapin seemed to be in a good mood when he joined Sam and Pete the next day.

```
Accounting


Data Prep
Service


Engineers


            A: 2 days                                    D: 2 days
Pete        ------------------------------:              -----------------------:


            A: 2 days                  B: 2 days         D: 2 days
Betsy       ------------------------------:-----------------------:-----------------------:


                                       C: 2 days
Sam                                    ------------------------------:

Days        --+--+--+--+--:--+--+--+--+--1--+--+--+--+--:--+--+--+--+--2--+--+--+--+--:--+
```

A: Prototype training   G: Develop Monthly Budgets
   Build initial prototype

B: Install network station   H: Standardize & proof
   & network dBASE III PLUS   job cost entries

C: Test prototype   I: Ad hoc reporting training

D: Revise prototype   J: Key job cost ledger
                                            entries to diskette

E: Develop ad hoc training   K: Develop accounting system
                                            data entry program

F: Integrate automated   L: Test data entry program
   & manual procedures

                                         M: Build & test file transfers

FIGURE 8.4

**Gantt chart for Buy/Build Software Phase of
cost control system (Part 1 of 3).**

```
                                        K: 5 days
                                        ------------------------------------------------:

                                                    J: 1 day
                                                    --:

            C: 2 days
            -----------------------------------:

E: 2 days                                           H: 2 days
-----------------------------------------:          -----------------:

                        D: 2 days            .      H: 2 days           D: 2 days
                        -----------------------:-----------------:-----------------:

C: 2 days                               G: 2 days        C: 2 days
-----------------------------------------:-----------------:-----------------:
--+--+--+--3--+--+--+--+--+--:--+--+--+--+--4--+--+--+--+--+--:--+--+--+--+--5--+--+--+--+--+--:--+
```

A: Prototype training   G: Develop Monthly Budgets
   Build initial prototype

B: Install network station   H: Standardize & proof
   & network dBASE III PLUS   job cost entries

C: Test prototype   I: Ad hoc reporting training

D: Revise prototype   J: Key job cost ledger
                                                        entries to diskette

E: Develop ad hoc training   K: Develop accounting system
                                                        data entry program

F: Integrate automated   L: Test data entry program
   & manual procedures

                                                    M: Build & test file transfers

FIGURE 8.4

**Gantt chart for Buy/Build Software Phase
for cost control system (part 2 of 3).**

```
                                                                Accounting

                                                                Data Prep
                                                                 Service


                                                                Engineers

I: 2.5 days
---------------------------------------------------------------------:        Pete


C:,  L:,  M: 3 days, I: 3.5 days                     F: 1 day    D: 2 days
----------------------------------------------------:-----:----------------: Betsy


                              C: 2 days        F: 1 day
                              ----------------:-----:              Sam
--+--+--+--6--+--+--+--+--+--:--+--+--+--+--+--7--+--+--+--+--:--+--+--+--+--+--8--+--+--+ Days
```

**Gantt chart for Buy/Build Software Phase
for cost control system (part 3 of 3).**

Mr. Chapin began, "I think you and Pete did a thorough job in your analysis, but I have some questions about your evaluation of the new system benefits. How soon will we begin to realize the $125,000 benefit under Job Cost Reports?"

"We had three engineers leave us last year," said Sam. "We can expect the same this year. Depending on the person, I will either make a counter offer, replace him or her, or absorb the workload into the remaining staff as time is freed up by the new system. I would like to absorb one position this year and one next year."

"That's a big commitment on your part, Sam, but I think it can be done," Mr. Chapin said. He continued, "What about the Monthly Budget benefits?"

Sam said, "The same two-year time frame should be enough, but we'll probably see very little this year and the bulk next year."

"How long will it take to get the system running?" asked Mr. Chapin.

"We're estimating four months," said Sam. "That's based upon the average length of a prototyping project and a maximum commitment of one-third of Betsy's time or my time at any given moment. The prototyping process is unpredictable, so I will keep you posted."

"Based upon that schedule, I approve your proposal to proceed with the project," Mr. Chapin said. "I'll look for the system to be running in four months. At the end of twelve months, we'll look at the staff in terms of size and turnover, and we'll review budget vs. actual and vendor experience. I'll be very happy if we achieve the results you indicated, Sam. However, the real benefit of your proposal lies in the start of a company-wide job cost/ budgeting project. We've suffered without one, but everytime we sent out requests for bids, the prices and the uncertainty about our needs were so high, we refused to make the commitment. I agree with you, Sam, when you say we need more than a job cost accounting system to manage our operations effectively, but no one here can say exactly what we should do.

"Your proposal allows us to get started, and it pays for itself in a short time. I am very pleased, and I want to be involved. I want the other department managers to follow your progress also. You've already included Ed Henderson from Accounting in your work. I'll help with the others. There will be some resistance, but I will make it clear that as research and development, I see this project as a top priority for the company. Any questions?"

Sam and Pete had no questions, so Mr. Chapin returned to his office. When Sam and Pete were alone, Pete asked, "How do you feel?"

"I don't know," said Sam. "He certainly doesn't waste much time. We'd better get to work."

Beginning the Design
of the New System

Sam and Pete worked for two hours after their meeting with Mr. Chapin. Pete explained the Design Phase of the systems development life cycle. "Throughout this phase, we'll consider the five components of a computer-based information system [3]: hardware, software, data, procedures, and personnel," he said. "We'll use the final Designer's Tradeoff Chart, Figure 7.4, as a detailed guide to the five components. We'll also use the Design DFD, Figure 6.11. You should make careful notes of this session, Sam.

"For hardware, we'll consider ACCESS, FLEXIBILITY OF OUTPUT, TURNAROUND TIME ON OUTPUT, and TIMELINESS OF DATA from the tradeoff chart. We're fortunate the network is in place already, and there is room for the fourth engineer's station in the current work area. You and Betsy will keep your stations, so there is very little to do to meet the specifications of the Tradeoff Chart as far as hardware is concerned."

"That's fine," said Sam. "No one has asked for a station at his or her desk yet. So, adding the fourth station to the existing cluster makes sense. I'll order it today."

"It is difficult to separate personnel from procedures, but let's start with personnel," said Pete. "We'll follow the tradeoff chart from SYSTEM RESPONSIBILITIES through LEARNING, then we'll loop back through the OUTPUT section.

"Look at SYSTEM RESPONSIBILITIES. Betsy has agreed to act as system manager, so she gets responsibility for the software, which means she fixes any bugs in the programs and makes changes to the programs as needed. She'll also be responsible for purchasing, installing, and maintaining the dBASE III PLUS software on the network."

"As far as responsibility for hardware goes, we use an outside maintenance company, and I'm satisfied with them," said Sam. "I'll let them know about the additional network station."

"For DATA RESPONSIBILITIES we'll need the Design DFD, Figure 6.11. It shows two automated data stores, EXPENSES and BUDGETS, which need to be maintained. That corresponds to responsibility for source documents, data entry, and updating on the Designer's Tradeoff Chart. Suppose Betsy Klein takes responsibility for EXPENSES and you take responsibility for BUDGETS."

"That seems logical," said Sam. "Betsy posts job cost ledger entries now, but I've never worked with a computer system in this way. What's involved?"

"Basically, you'll be responsible for timely and accurate data entry. Further details will depend upon what we come up with for the database design. You'll also have to maintain the **source documents** to resolve discrepancies and to satisfy any audit requirements."

"Will it involve much time?" asked Sam.

"You have at most ten jobs going at any one time, and you deal with four general ledger accounts," said Pete. "So you have to make forty budget entries per month. The time for data entry and maintenance will be minimal compared to the time spent managing the budgets. You could prepare the data on a form and let Betsy enter it, but I don't think you gain anything and that introduces extra communication requirements."

"OK," said Sam.

"The rest of the personnel assignments are straightforward," said Pete. "Data conversion will be Betsy's responsibility. We'll know more about what's involved when we're done with the database design and the initial prototype."

"Betsy also gets responsibility for **backup** and **restore**, which involves making backup copies of the system programs and data in case the working system is destroyed. In the event of a loss, Betsy will use the backup copies to restore the system, and determine what data, if any, needs to be reentered. Right now, she is backing up the word processing system three times per week. Since expenses will be entered every day, I recommend a daily backup for this system. It should take only about five minutes.

"Continuing down the tradeoff chart, you can see Betsy will be responsible for keeping your system current with the accounting system. The easiest thing to do is to produce a year-to-date general ledger Budget vs. Actual Report and compare it each month to the Accounting System Report. If there is a discrepancy, then she'll have to compare the source documents to the reports to uncover the error. Since we'll eventually do our data entry through the accounting system, this will be a temporary responsibility.

"The next item on the tradeoff chart is security and confidentiality. This is a problem in a network environment. Is any information kept from Betsy?"

"No," said Sam. "She needs to see everything, and she knows everything. I trust her."

"Good," said Pete. "We'll assign responsibility to her. We talked about the confidentiality of salary information during system analysis. Is there anything else?"

"No," said Sam.

"The last responsibility for Betsy on the tradeoff chart involves the training program," said Pete. "I will help her design it, and she will deliver it. You will have to meet with her to decide which of her current duties she will give up to take on the system manager's job. Those duties will have to be reassigned. You also need to discuss compensation with her."

"I'll take care of that," said Sam.

"Then, since everyone is responsible for his or her own output, we are finished with personnel involved with the automated system," said Pete. "Before we move to procedures, we should discuss the vendor control portion of the system.

"You should compile a list of the products and services you purchase from vendors, Sam. Organize the list according to whatever method you see fit. Use your word processor and spreadsheet program to help with the organization."

"I've already started," said Sam. "I'm using the word processor, but I did not think of using the spreadsheet. Thanks."

"Let's move to procedures," said Pete. "We talked about some procedures when we were discussing personnel. How is Betsy's accuracy with the job cost ledger entries?"

"Fine," said Sam.

"Then we'll plan to adapt her posting procedure to both EXPENSES and BUDGETS data entry," said Pete. "We've already discussed backup and controlling the match with the accounting system. That's enough for now. Of course, you will continue to perform all of the manual procedures of the Analysis DFD that were not chosen for automation. In the budget we set aside one day for you and Betsy to rework those procedures and integrate them with the automated system. I agree with your decision to schedule this activity near the completion of the fourth prototype. Since we're prototyping the software, the integration of the automated system with the manual should continue to evolve throughout the process.

"That brings us to the data and the software," continued Pete. "These components will evolve during the prototyping process. Since the design of hardware, procedures, and personnel yielded no surprises, Betsy and I will begin that work tomorrow. I've got a fair idea of your basic needs, and Betsy knows more. I think we'll have something for you to look at in about two days. The initial prototype will run as a stand-alone system while we're waiting to install the network version of dBASE III PLUS."

"I'm anxious to get my hands on a working system," said Sam. "I'm glad I have to wait only two days."

"That's the advantage of the prototyping method of software development," said Pete. "In a less sophisticated programming environment, systems specialists would be beginning a detailed design of the data and the software at this point. When the detailed design of the entire system is complete, programming costs are estimated and the feasibility of the entire project is reviewed. If the decision is in favor of proceeding, the development team agrees to "freeze" the design, and programming begins. When all the programs are written, the system is delivered to the users.

"With prototyping we design as we go along," continued Pete. "Since we're working in a fourth-generation/CASE environment, data and software designs are easily changed. We've done our analysis, and we basically know what you want. We'll put it all together in a working system and you'll experiment with it. If you find something you don't like, we'll change it or throw it away and start over again. We'll continue to evaluate and change the system until we arrive at a satisfactory product or until we decide to quit the project. Prototyping exploits the fourth-generation/CASE technology to encourage user participation and experimentation in the design and programming of the data and software components."

Summary

In the Design New System Phase of the systems development life cycle the development team must consider hardware, software, data, procedures, and personnel regardless of the choice of software development methodology. Since software development methodology affects the software and data components primarily, the team should consider hardware, procedures, and personnel first, although it will be impossible to separate software and data from these three completely. It will also be impossible to consider each component in isolation because choice of hardware affects procedures, which affects personnel, etc.

The choice of hardware environment was made during the Evaluation of Alternatives Phase of the life cycle. In the Design New System Phase the development team considers specifics, such as the physical layout of the hardware, to satisfy the requirements of the ACCESS through TIMELINESS OF DATA sections of the Designer's Tradeoff Chart.

In considering personnel, the development team first looks at the SYSTEM RESPONSIBILITIES specified in the tradeoff chart to assign system management functions. In moving to DATA RESPONSIBILITIES, the Design DFD is also used. The development team must assign responsibility for the maintenance of each data store on the Design DFD to specific personnel. When this is done, other data responsibilities such as backup and security, are assigned.

Training and output responsibilities are assigned next, as are secondary development efforts, such as the initiation of the vendor control database.

In considering procedures, the development team looks at the DATA RESPONSIBILITIES section of the tradeoff chart for day-to-day data maintenance activities. The TIMELINESS OF DATA section is also considered at this time. In the Horatio & Co. Cost Control System design each person generates his or her own output at a network workstation as needed, so no procedures for the preparation of output are required. In systems with less flexible output components, procedures for the processing of output would be designed from the specifications in the ACCESS, FLEXIBILITY OF OUTPUT, and TURNAROUND TIME ON OUTPUT sections of the tradeoff chart.

In discussing data and software design, the development team considers prewritten software packages, prototyping, and the traditional detailed design/programming method. If the required information system solution is available in a feasible software package, then data and software development is complete, and the project proceeds to the Implementation Phase of the systems development life cycle after the purchase of the software package.

If the required information systems solution is not available in a package, then it must be developed. The choice between prototyping and the traditional detailed design method is based upon the feasibility of a fourth-generation/CASE environment and the time commitment from the user/management group. If prototyping is feasible, then the analysis to date is sufficient to develop a working system satisfying the basic requirements of the Requirements Model. If the design of the hardware, procedures, and personnel components does not uncover any difficulties, the project can move to the Buy/Build Software Phase of the systems development life cycle.

If prototyping is not feasible, then the system programming requirements will be substantial. The development team should design and specify as much of the data and software components as possible before beginning programming. With the completion of the detailed design, the feasibility of the entire project—economic, technical, and operational—should be reviewed before proceeding to the Buy/Build Software Phase of the systems development life cycle.

THOUGHT QUESTIONS
In Your Opinion. . .

1. What do you think of Sam's decision to schedule part-time commitments to the cost control system? What do you think were his alternatives? Which alternative do you prefer? Why?

2. Sam brought his report to Mr. Chapin rather than sending it through the interoffice mail. Do you think how the report was delivered affected Mr. Chapin's opinion of it? Why or why not?

3. Do you think Mr. Chapin understood the uncertainty in the project schedule due to the use of prototyping? If you were Sam, how would you follow up?

4. Do you think many prototyping projects fall into an endless cycle of evaluation and revision? Why or why not?

5. What security problems are faced in a network environment? Are these any different than those faced in a mainframe computing environment? Are these any different than those faced with manual systems?

6. How do you think Sam should approach Betsy Klein regarding her new job responsibilities? How should Betsy approach Sam? Is Betsy justified in asking for an increase in compensation? If you were Sam, how would you react to such a request? If you were Mr. Chapin, how would you react?

References

1. Jenkins, A. M. "Prototyping: A Methodology for the Design and Development of Application Systems," *Spectrum*, Volume 2, Number 2, April 1985, pp. 1–8.

2. Jenkins, A. M. "Prototyping: A Methodology for the Design and Development of Application Systems—Part 2," *Spectrum*, Volume 2, Number 3, June 1985, pp. 1-4.

3. Kroenke, D., and K. Dolan. *Business Computer Systems,* Third Edition. Santa Cruz, CA: Mitchell Publishing Corporation, 1987.

4. Page-Jones, M. *Practical Project Management*. New York: Dorset House Publishing,1985.

HORATIO&CO.

Designing the Automated System

The exercises of this section ask you to repeat the processes described in this chapter. Since Camille Abelardo decided to use a prewritten software package, she is not concerned with the development of the software component of the credit union system. However, she is concerned with the remaining components of a computer-based information system, especially hardware, procedures, and personnel.

The exercises also show that Camille must be concerned with project scheduling and management, although these tasks are greatly simplified because of the decision to use a prewritten software package.

In the exercises of this chapter and beyond, you will develop a realistic assessment of the similarities and differences of prewritten software package projects vs. projects that develop the software component of the new system.

Exercises

1. Review the Statement of Business Objectives, Business Tactics, and System Objectives you prepared in Chapter 4. Expand the report with the specifics gathered during systems analysis.

2. Organize your work for exercise 7, Chapter 7, into a list of credit union system resource requirements. Use Figure 8.3 as a guide.

3. Organize the credit union system resource requirements developed in the previous exercise into a Gantt chart. Use Figure 8.4 as a guide.

4. Does Camille's decision to proceed to the Buy/Build Software Phase of the systems development life cycle seem reasonable? Discuss.

5. Design the hardware layout for the credit union system. Use the access, flexibility of output, turnaround time on output, and timeliness of data from the tradeoff chart you prepared for the Chapter 7 exercise.

6. Design the procedures and the personnel components of the credit union system from Chapter 7 tradeoff chart. Follow the tradeoff chart from system responsibilities through learning, then loop back through the output section.

7. Compare the design of hardware, procedures, and personnel for the cost control and credit union systems. Discuss similarities and differences.

CHAPTER 9

Building the Software

Objectives

In Chapters 2 and 3 you used the initial prototype of the cost control system. In this chapter you will observe the development of the database and the programs of that prototype.

The specific objectives of this chapter are

1. To analyze the output data flows of the Design DFD to develop preliminary data store contents;
2. To develop a database design from the contents of the data stores;
3. To determine the required access paths into the database;
4. To represent the relationships within the database with a data structure diagram;
5. To determine appropriate system processes;
6. To represent the organization of the system processes with a system structure chart.

The Data Component: Output Data Flows

Pete and Betsy began their scheduled meeting on time. They agreed to work as a team on the data and software components instead of dividing the work between them.

Pete began, "The purpose of this session is to determine the contents of the data flows and the data stores on the Design DFD." (Figure 6.11, repeated as Figure 9.1.) "One message came through loud and clear during my interviews with Sam and the engineers. The job cost ledger is vital to the operation of the department. Let's begin with the contents of the output data flow marked Job Cost Reports."

Job Cost Reports

"The simplest is the Monthly Job Cost Report, which goes to the project managers," said Betsy. "It lists the job number and the total expenditures made by this department for the month." Pete retrieved Figure 5.4 from his file and a supply of forms marked DATA FLOW/DATA STORE CONTENTS. He made the entries shown in Figure 9.2.

"The **contents form** is very much like a DATA DIVISION entry in a COBOL program," said Pete. "At the end of our work we will have a contents form for each data flow and data store. This collection of contents forms will comprise our **data dictionary**. The data dictionary helps to organize and document the development of a new system."

"I understand," said Betsy. "The top part of the form is clear. The body of the form shows the contents of the data flow broken down into **components**. Some components are indented to show that they are subcomponents of a component listed above. Job Number and Amount are subcomponents of the component named Monthly Job Cost Report Line.

"The entry in the **repetitions** column shows that there can be any number of lines on the report, each containing a job number and an amount, while Department Name, Month–Year, and Grand Total appear once at the top or bottom of the report. How the repetitions occur is explained in the Comments column." See Figure 5.4 for the actual report and compare it to the contents form, Figure 9.2.

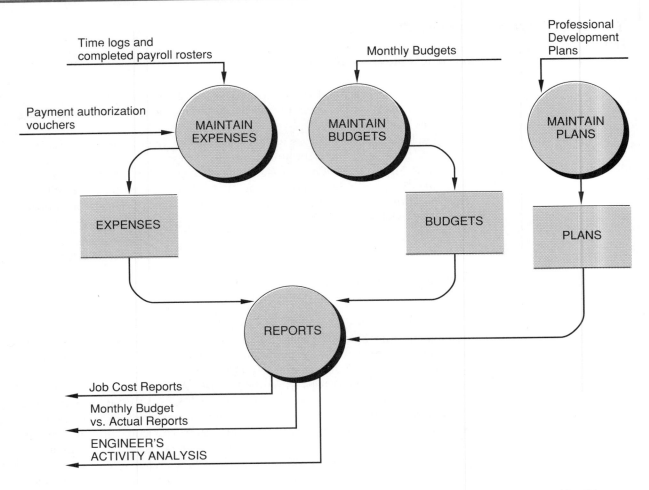

Time logs and
completed payroll rosters

Payment authorization
vouchers

MAINTAIN
EXPENSES

Monthly Budgets

MAINTAIN
BUDGETS

Professional
Development
Plans

MAINTAIN
PLANS

EXPENSES

BUDGETS

PLANS

REPORTS

Job Cost Reports

Monthly Budget
vs. Actual Reports

ENGINEER'S
ACTIVITY ANALYSIS

EXPENSES = Job cost ledger + time logs

PLANS = Semiannual by engineer

BUDGETS =
Monthly
by Job
by Account

Note: Vendor Control?

FIGURE 9.1

**The Design DFD (repeated from
Figure 6.11).**

DATA FLOW CONTENTS		
Name: Monthly Job Cost Report		
Repetitions	*Component*	*Comments*
	Department Name	
	Month–Year	
1–?	Monthly Job Cost Report Line	One for each job
	Job Number	
	Amount	Total expenditures for month
	Grand Total	

FIGURE 9.2

**Contents of Monthly Job Cost Report
data flow.**

Betsy continued, "We also provide Project-To-Date Job Cost Reports to the project managers. The format is the same except that Amount represents the project-to-date expenditures through the month of the report for the given job." Betsy completed a contents form for the year-to-date report. It is shown in Figure 9.3.

"During my talks with Sam, he mentioned requests for such information as what did we spend on engineering for Job C722," said Pete.

"That comes up all the time," said Betsy. "What did we spend on a general ledger account for a certain job is the most common situation. We also get what did we spend on a certain job, what did we spend in a certain account, and what did we spend in a certain month. Sometimes they just want the totals, but most of the time they want to see the individual expenses to get a clear picture of what happened, and they want it right away. A day or two later is useless."

"How about an inquiry that would allow a person to select data from the job cost ledger by month, by source, by account, by job, and by job and account?" asked Pete.

"That would be fine," said Betsy. "I'll make up the contents form." Betsy's form for the inquiry is shown in Figure 9.4.

DATA FLOW CONTENTS

Name: Project-to-Date Job Cost Report

Repetitions	*Component*	*Comments*
	Department Name	
	Month–Year	
1–?	PTD Job Cost Report Line	One for each job
	Job Number	
	Amount	PTD expenditures thru month
	Grand Total	

FIGURE 9.3

Contents of Project-to-Date Job Cost Report data flow.

DATA FLOW CONTENTS

Name: Inquiry

Repetitions	*Component*	*Comments*	
	Date		
	Source		
	Account Number		
	Account Name		
	Job Number		
	Amount		
		Selection	by month
			by source
			by account
			by job
			by job & account

FIGURE 9.4

Contents of Inquiry.

Budget vs. Actual Reports

"Are there any other Job Cost Reports or inquiries?" asked Pete.

"None that I can think of," said Betsy.

"I'm sure something will come up later," said Pete. "Let's move to the Monthly Budget vs. Actual Report data flow on the Design DFD. "Sam and I thought that the system should provide Budget vs. Actual Reports by general ledger account across all jobs and for each job individually. The reports would be for a single month or for expenses-to-date."

"That's four reports," said Betsy. "What would they look like?"

Pete took a clean contents form. "Most Budget vs. Actual Reports rely upon a cutoff date. Only expenses incurred on or before the cutoff date are included in the Actual Expenses column of the report. The report for an individual job would show budget and actual expense figures for that job, either from the first of the month containing the cutoff date through the cutoff date or from the start of the project through the cutoff date. The total report would consider only the general ledger accounts without regard to jobs."

"We get the total report from the accounting system now," said Betsy. "Why do we need to repeat it?"

"This report will serve as a control on the match between the data entered into both systems," said Pete.

"That's good," said Betsy. "I was wondering how that would be done."

Expense History Reports

"Since I'll be responsible for maintaining the accuracy of the data, let me add an output of my own," said Betsy.

"Sure," said Pete. "What is it?"

"I want to be able to print out a history of all expenses by account number and month. I'll use this report to resolve any differences with the accounting system. While we're at it, we may as well include a history report by job; it will help with the job closeout procedures.

"I do not want to have to key in which account or which job I want," continued Betsy. "I am not talking about a management report. This report is for accuracy control. I want all accounts or jobs to come out automatically."

Contents forms for the Monthly Job Budget vs. Actual Report, the Project-To-Date Job Budget vs. Actual Report, the Monthly General Ledger (G/L) Budget vs. Actual Report, and the Year-To-Date General Ledger (G/L) Budget vs. Actual Report are shown in Figures 9.5 through 9.8, respectively.

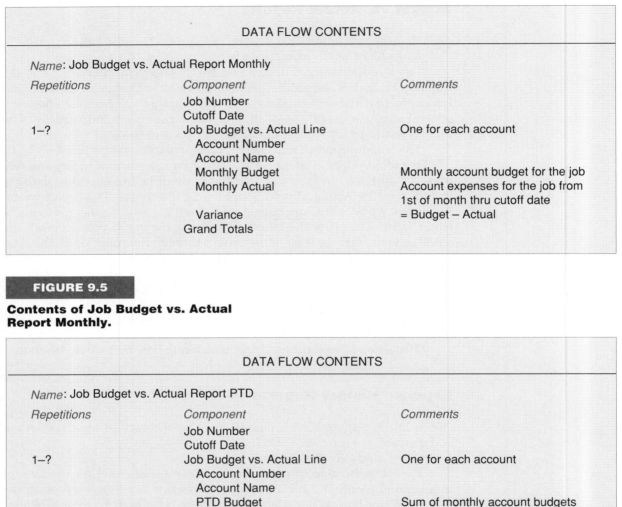

DATA FLOW CONTENTS

Name: Job Budget vs. Actual Report Monthly

Repetitions	*Component*	*Comments*
	Job Number	
	Cutoff Date	
1–?	Job Budget vs. Actual Line	One for each account
	Account Number	
	Account Name	
	Monthly Budget	Monthly account budget for the job
	Monthly Actual	Account expenses for the job from 1st of month thru cutoff date
	Variance	= Budget – Actual
	Grand Totals	

FIGURE 9.5

**Contents of Job Budget vs. Actual
Report Monthly.**

DATA FLOW CONTENTS

Name: Job Budget vs. Actual Report PTD

Repetitions	*Component*	*Comments*
	Job Number	
	Cutoff Date	
1–?	Job Budget vs. Actual Line	One for each account
	Account Number	
	Account Name	
	PTD Budget	Sum of monthly account budgets for the job from start thru month of cutoff date
	PTD Actual	Account expenses for the job from start thru cutoff date
	Variance	= Budget – Actual
	Grand Totals	

FIGURE 9.6

**Contents of Job Budget vs. Actual
Report PTD.**

DATA FLOW CONTENTS

Name: G/L Budget vs. Actual Report Monthly

Repetitions	Component	Comments
	Cutoff Date	
1–?	G/L Budget vs. Actual Line	One for each account
	Account Number	
	Account Name	
	Monthly Budget	Monthly budget for the account
	Monthly Actual	Account expenses for the job from 1st of month thru cutoff date
	Variance	= Budget − Actual
	Grand Totals	

FIGURE 9.7

Contents of G/L Budget vs. Actual Report Monthly.

DATA FLOW CONTENTS

Name: G/L Budget vs. Actual Report YTD

Repetitions	Component	Comments
	Cutoff Date	
1–?	G/L Budget vs. Actual Line	One for each account
	Account Number	
	Account Name	
	YTD Budget	Sum of monthly account budgets from 1st of year thru month of cutoff date
	YTD Actual	Account expenses for the job from 1st of year thru cutoff date
	Variance	= Budget − Actual
	Grand Totals	

FIGURE 9.8

Contents of G/L Budget vs. Actual Report YTD.

The contents of Betsy's history reports are shown in Figures 9.9 and 9.10. Compare the contents form in Figure 9.5 to the Job Budget vs. Actual Report shown in Figure 2.5. Compare the contents forms in Figures 9.9 and 9.10 to the Expense History Reports shown in Figures 2.9 and 2.10. Notice that the Job Budget vs. Actual Report represents one job, while the Expense History Report by Job presents all jobs automatically, and notice the resulting difference in the contents forms for each report, Figures 9.5 and 9.10.

Engineers' Activity Analysis

"The last output data flow to consider is the ENGINEERS' ACTIVITY ANALYSIS," said Pete. "My notes say Sam needs an up-to-the-minute, accurate tabulation of each engineer's activity to see exactly how the engineer

DATA FLOW CONTENTS

Name: Expense History Report by Account

Repetitions	*Component*	*Comments*
1–?	Account Section	One for each account
	Account Number	
	Account Name	
1–?	Month Section	One for each month within account
	Month	
1–?	Expense History Report Line	One for each expense within month
	Date	
	Source	
	Description	
	Job Number	
	Amount	
	Subtotal by Month	
	Subtoal by Account	
	Grand Total	

FIGURE 9.9

Contents of Expense History Report by Account.

DATA FLOW CONTENTS

Name: Expense History Report by Job

Repetitions	*Component*	*Comments*
1–?	Job Section	One for each job
	Job Number	
1–?	Month Section	One for each month within job
	Month	
1–?	Expense History Report Line	One for each expense within month
	Date	
	Source	
	Account Number	
	Account Name	
	Description	
	Amount	
	Subtotal by Month	
	Subtotal by Job	
	Grand Total	

FIGURE 9.10

Contents of Expense History Report by Job.

spent his or her time. Sam indicated a preference for an inquiry on this one. Will the inquiry we've already designed (Figure 9.4) work?"

"I don't think so," said Betsy. "The engineer's name is in the Source component, so we could select by source, but there is no indication of how much work was done or what kind of work was done, only the cost of the time."

"The hours and the descriptions of the tasks are on the time logs, but they do not get entered into the job cost ledger," said Pete. "We could add a description component and an hours component to the contents of Inquiry. This would allow Sam to make the judgements he wants, but the format is difficult. Let's add a special ENGINEERS' ACTIVITY ANALYSIS INQUIRY."

The modified version of the Inquiry contents form is shown in Figure 9.11. The contents form for the ENGINEERS' ACTIVITY ANALYSIS is shown in Figure 9.12.

The work on the ENGINEERS' ACTIVITY ANALYSIS completed the analysis of the output data flows of the Design DFD. Pete and Betsy will use this work to determine the contents of the Design DFD's data stores next.

DATA FLOW CONTENTS

Name: Inquiry

Repetitions	*Component*	*Comments*
	Date	
	Source	
	Account Number	
	Account Name	
	Job Number	
	Description	
	Amount	
	Hours	Selection by month
		by source
		by account
		by job
		by job & account

FIGURE 9.11

Modified contents of screen inquiry.

The Data Component: Data Stores

The Design DFD, Figure 9.1, shows two automated data stores: EXPENSES and BUDGETS. The contents of Design DFD data stores are developed using the principle that all data items appearing in the Design DFD output data flows must come from the Design DFD data stores.

The programming environment in which the application software is developed affects the choices made at this step of the Design New System phase. At this point it is useful to express data store design concepts in the language of the programming environment.

Like many other environments, dBASE III PLUS deals with files of records that are made up of fields. Direct access to the records of a dBASE III PLUS file may be made through the use of a file index. dBASE III PLUS allows files that share a common field to be connected to each other through a mechanism called a file relation.

Indexed files of records are available in almost all programming environments. File relations are found only in database management system-

DATA FLOW CONTENTS

Name: ENGINEERS' ACTIVITY ANALYSIS

Repetitions	*Component*	*Comments*
	Engineer	
1–?	Activity Section	One for each activity within engineer
	Activity	
1–?	Activity Analysis Line	
	Date	
	Source	
	Description	
	Job	
	Hours	
	% of total Hours	
	Totals by Activity	
	Grand Total	
		Selection by engineer

FIGURE 9.12

Contents of ENGINEERS' ACTIVITY ANALYSIS.

like environments. The concepts used here are dBASE III PLUS specific, but they easily transfer to similar environments.

Developing the contents of the Design DFD data stores involves the systematic review of the contents forms for the Design DFD output data flows. Four rules, summarized in Figure 9.13, govern the review of the forms.

The EXPENSES Data Store

The Inquiry output data flow, Figure 9.11, satisfies rule 4, so it is considered first. This data flow is logically associated with the EXPENSES data store because in the Design DFD, Figure 9.1, EXPENSES logically replaces the manual job cost ledger and the manual time logs, which are the sources of

1. All items that appear on the output data flows must appear in a data store. Exception: items, such as cutoff date for the Budget vs. Actual Reports, that are entered by the user at the time the output is produced.

2. In addition to the components that comprise the contents of the output data flow under consideration, the data store must contain the components that identify records for selection, sequencing, and subtotals.

3. Items that can be computed from other items in a data store should not be included in the data store.

4. Begin the review, if possible, with an output data flow that can be associated with a single data store.

FIGURE 9.13

Rules for developing the contents of the Design DFD data stores.

the inquiries in the current system. Each record of the EXPENSES data store must contain fields for the date, source, account number, account name, job number, description, amount, and hours of the expense in order to select and display information for the Inquiry. The EXPENSES data store components are recorded on the contents form shown in Figure 9.14.

The Expense History Reports, Figures 9.9 and 9.10, are also associated with the EXPENSES data store, so they are reviewed next. The rules of Figure 9.13 govern the review.

No new components are contained in the reports. No record selection is required, and the sequencing is either by account or job. Therefore the EXPENSES data store contents form remains unchanged.

The format of the Expense History Reports is much more complicated than that of the Inquiry. The format, known as a control-break report, is typical of transaction processing and/or reporting system outputs. The transformation of the EXPENSES data store records into the Expense History Reports will be performed by the Expense History Report programs.

The Monthly Job Cost Report and the Project-To-Date Job Cost Report, Figures 9.2 and 9.3, are both associated with EXPENSES. Total monthly and project-to-date expenditures for a job can be computed by totaling the Amount field of the EXPENSES data store records by job number. Since the sequencing and the identification of the proper records and the required computations can be performed with the existing components, no new entries on Figure 9.14 are required.

DATA STORE CONTENTS

Name: EXPENSES

Repetitions	*Component*	*Comments*
	Date	
	Account Number	
	Account Name	
	Job Number	
	Source	
	Description	
	Amount	
	Hours	

FIGURE 9.14

Preliminary contents of EXPENSES data store.

The Month–Year component of Figures 9.2 and 9.3 would normally be entered by the user at the time the report is run, so no entry on Figure 9.14 is necessary for this component either.

The Department Name component of Figures 9.2 and 9.3 is an example of **control information**. All computer-based information systems require some sort of control information. Control information is usually kept in a separate data store. Figure 9.15 represents the contents form for the CONTROL data store. No entry in Figure 9.14 is necessary for the Department Name component.

The remaining output data flows correspond to the Budget vs. Actual Reports. For the reports for individual jobs the user will enter the job number and the cutoff date at the time the report is run. For the reports across all jobs the user will enter the cutoff date at the time the report is run. No data store notations are necessary for these items.

For the Job Budget vs. Actual Report Monthly, Figure 9.5, the Monthly Actual component is computed from the EXPENSES data store. Using only the EXPENSES data store records that correspond to the requested job number and cutoff date, the Amount field of the EXPENSES data store records is totaled by account. Since Job Number, Date, Account Number, and Amount are already listed on Figure 9.14, no additional notations are necessary.

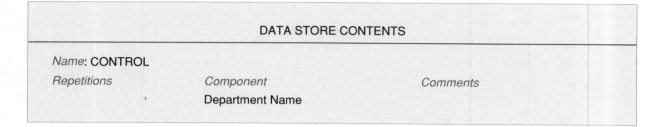

DATA STORE CONTENTS

Name: CONTROL

Repetitions	*Component*	*Comments*
'	Department Name	

FIGURE 9.15

Preliminary contents of CONTROL data store.

The PTD Actual component of the Job Budget vs. Actual Report PTD, Figure 9.6, is computed similarly. The Actual components of the G/L Budget vs. Actual Reports, Figures 9.7 and 9.8, follow the same procedure but ignore job numbers. Again the original entries on Figure 9.14 remain unchanged.

The BUDGETS Data Store

For the Job Budget vs. Actual Report Monthly, Figure 9.5, the Monthly Budget component cannot come from the EXPENSES data store. To produce the report in Figure 9.5, the BUDGETS data store must contain the monthly budget figures identified by an account/job combination. The contents form for the BUDGETS data store appears in Figure 9.16.

The PTD Budget component of the Job Budget vs. Actual Report PTD, Figure 9.6, is computed by totaling the monthly budget amounts for the proper account/job combination.

The Monthly G/L Budget vs. Actual Report (Figure 9.7) program will combine all BUDGETS data store records for a given account to produce the Monthly Budget for the account across all jobs. The YTD G/L Budget vs. Actual Report (Figure 9.8) program will perform a similar combination. The contents form shown in Figure 9.16 is sufficient for these calculations.

DATA STORE CONTENTS

Name: BUDGETS

Repetitions	*Component*	*Comments*
	Account Number	
	Account Name	
	Job Number	
1–12	Monthly Budget Amount	

FIGURE 9.16

Preliminary contents of BUDGETS data store.

Designing the Database from the Data Stores

Data stores represent collections of data on data flow diagrams. In dBASE III PLUS systems, data stores often become files in the final physical implementation of the Design DFD; sometimes more than one file is required to implement the contents of a data store properly.

Step 1: Identify Master and Transaction Data Stores

Database design for the cost control system begins with a comparison of the Design DFD, Figure 9.1, with the data flow diagram for a general transaction processing and/or reporting system, Figure 3.16. The purpose of the comparison is to identify the master and transaction data stores of the cost control system. Figure 3.16 is repeated as Figure 9.17.

In transaction processing and/or reporting systems, **master data stores** contain background information on such system entities as the name of a customer or the price of an inventory item. **Transaction data stores** are used to capture the day-to-day activity of the system.

In the cost control system the BUDGETS data store represents **background information** on general ledger accounts and jobs (master), while the EXPENSES data store represents **day-to-day activity** against the accounts and the jobs (transaction).

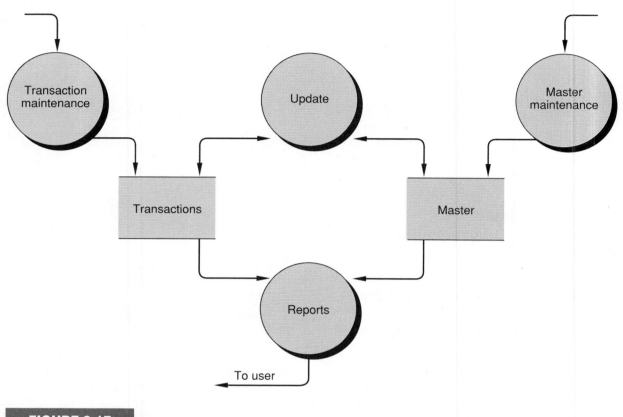

FIGURE 9.17

Data Flow Diagram for a transaction processing system (repeated from Figure 3.16).

Step 2: Identify the Data Store Keys

The second step in database design is the identification of the **key** of each data store. This is accomplished through a review of the data store contents forms, Figures 9.14 and 9.16.

A data store key is the component or components that uniquely identify a given record. In the BUDGETS data store, Figure 9.16, the combination of Account Number and Job Number uniquely identify a record, so the

Account Number/Job Number combination is the key to the BUDGETS data store.

Account Number alone is not enough to identify a BUDGETS data store record uniquely because, for a given account, the data store contains a budget record for each job. Similarly, Job Number alone is not enough to identify a record uniquely. Only when both an Account Number and a Job Number are given can one find a single matching BUDGETS data store record.

The key to the EXPENSES data store is more complicated. Many expense records exist for each Account Number/Job Number combination, so more components must be added to the key. The Date field helps identify an expense record, but an Account Number/Job Number combination can have more than one expense record for a given date. The same is true for Source. Finally, when one considers Date, Account Number, Job Number, Source, and Description, one has a complete identification of an EXPENSES data store record.

Step 3: Remove Redundant Nonkey Components

The third step in database design is the elimination of **redundant nonkey components**, that is components that appear in more than one data store and that are not part of the key of any data store. In EXPENSES and BUDGETS, Figures 9.14 and 9.16, Account Name is a redundant nonkey component. Since Account Name connotes background information, it should be left in BUDGETS, the master data store, and removed from EXPENSES, the transaction data store. The results of the first three steps of the database design process are shown in Figures 9.18 and 9.19.

Step 4: Remove Repeating Components

The fourth step in database design is the removal of **repeating components**. Repeating components are removed by determining the data necessary to identify an individual repetition uniquely. In BUDGETS, Figure 9.18, the month ending date determines an individual Monthly Budget Amount uniquely.

DATA STORE CONTENTS

Name: BUDGETS

Repetitions	*Component*	*Comments*
	<u>Account Number</u>	
	<u>Job Number</u>	
	Account Name	
1–?	Monthly Budget Amount	

underline = key component

FIGURE 9.18

BUDGETS data store with key.

DATA STORE CONTENTS

Name: EXPENSES

Repetitions	*Component*	*Comments*
	<u>Date</u>	
	<u>Account Number</u>	
	<u>Job Number</u>	
	<u>Source</u>	
	<u>Description</u>	
	Amount	

underline = key component

FIGURE 9.19

EXPENSES data store with key.

DATA STORE CONTENTS

Name: BUDGETS

Repetitions	*Component*	*Comments*
	<u>Account Number</u>	
	<u>Job Number</u>	
	Account Name	
	<u>Month Ending Date</u>	
	Budget Amount	

underline = key component

FIGURE 9.20

BUDGETS data store minus repeating component.

Once the unique identification data is determined, the database designer has two choices. The identification data can be added to the key of the data store, thereby replacing a long data store record containing a repeating component with many shorter records, one for each of the individual repetitions.

This was done for BUDGETS. The results are shown in Figure 9.20. One long record containing several monthly budget figures has been replaced by several shorter records containing a single budget figure along with the month to which the budget figure corresponds.

The second choice for the removal of a repeating component involves splitting the data store into two files. The first file contains the components of the original data store minus the repeating group. The individual repetitions of the repeating group are stored as records in the second file. The database designer must determine the key of the second file and add components to the contents form if necessary. In addition, the database designer must add components to the second file that enable programs to find the corresponding record in the first file, and vice versa.

Figures 9.21 through 9.23 illustrate this technique for a data store that might be used in a university registrar's office.

DATA STORE CONTENTS

Name: STUDENTS

Repetitions	*Component*	*Comments*
	<u>Social Security No.</u>	
	Name	
	Home Address	
	Home Phone	
	Campus Address	
	Campus Phone	
	Major	
	Credits Completed	
	Grade Point Average	
1–?	Completed Courses	
	Course Number	
	Semester	
	Year	
	Final Grade	

underline = key component

FIGURE 9.21

Data store containing a repeating component.

Step 5: Remove Partial and Transitive Dependencies

The fifth and final step in the database design is the removal of components that do not depend upon the entire key **(partial dependency)** or that depend upon nonkey components **(transitive dependency)**. In BUDGETS, Figure 9.20, Account Name represents a partial dependency. Account Name depends upon Account Number only. The remaining key components, Job Number and Month, do not affect Account Name. Components such as these should be moved to a separate data store, and the necessary key components should be added to the contents of the new data store. Figures 9.24 and 9.25 illustrate the technique for the BUDGETS data store.

DATA STORE CONTENTS

Name: STUDENT MASTER

Repetitions	*Component*	*Comments*
	<u>Social Security No.</u>	
	Name	
	Home Address	
	Home Phone	
	Campus Address	
	Campus Phone	
	Major	
	Credits Completed	
	Grade Point Average	

underline = key component

FIGURE 9.22

Original data store minus repeating component.

DATA STORE CONTENTS

Name: COURSES

Repetitions	*Component*	*Comments*
	<u>Social Security No.</u>	
	<u>Course Number</u>	
	<u>Semester</u>	
	<u>Year</u>	
	Final Grade	

underline = key component

FIGURE 9.23

Repeating component data in a separate data store.

DATA STORE CONTENTS

Name: ID

Repetitions	*Component*	*Comments*
	Account Number	
	Account Name	

underline = key component

Separate data store for Account Names; ID is part of the BUDGETS data store.

DATA STORE CONTENTS

Name: BUDGET

Repetitions	*Component*	*Comments*
	Account Number	
	Job Number	
	Month Ending Date	
	Budget Amount	

underline = key component

BUDGETS data store minus Account Name; BUDGET is just part of the BUDGETS data store.

The five steps of database design for a transaction processing and/or reporting system are summarized in Figure 9.26. The process is known as **normalizing the database**. The rules presented here are not complete [2], but they are sufficient for our purposes.

1. Identify Design DFD data stores as either master or transaction.

2. Identify the key to each data store.

3. Eliminate redundant nonkey components.

4. Eliminate repeating components.

5. Remove components that do not depend upon the key or that depend upon nonkey components.

FIGURE 9.26

Rules for transaction processing and/or reporting database design.

Determining Access Paths

Access paths represent the ways in which the system's programs access the data in the system's files. In dBASE III PLUS systems, access paths are implemented through file indexes. A separate file index must be established for every access path required by the system.

The unique identification keys identified in the contents forms for EXPENSES, ID, and BUDGET, Figures 9.19, 9.24, and 9.25, represent the primary access paths for the cost control system files. An index file corresponding to the unique identification key of each data file must be established.

Report and Inquiry sequences also dictate access paths. The Budget vs. Actual Reports require EXPENSES and BUDGET file totals by Account Number, so both BUDGET and EXPENSES must be accessed by Account Number. Since indexes are kept in files separate from the data file and other indexes, it is possible to create more than one index for a data file. At this point EXPENSES has two indexes: one by the unique identification key, Date+Account Number+Job Number+Source+Description, and the other by Account Number alone. BUDGET has one index by the unique identification key, Account Number+Job Number+Month, and ID has one index by the unique identification key, Account Number.

The Job Cost Reports require access to EXPENSES in Job Number sequence, so a third index for EXPENSES, by Job Number, has been identified. The Expense History Report by Job also requires access to EXPENSES in Job Number sequence.

Since the Expense History Report by Job requires subtotals by Month within Job, sequencing by Date within Job is necessary. Expanding the third index on EXPENSES to Job Number+Date satisfies this requirement while maintaining the sequence required by the Job Cost Reports.

The Expense History Report by Account requires access to EXPENSES in Account Number sequence which is provided by the second EXPENSES index file. Expanding the second index on EXPENSES to Account Number+Date provides for subtotals by Month while maintaining the sequence, required by the Budget vs. Actual Reports.

Inquiries use the EXPENSES file only. In the initial prototype, selected inquiry records will be presented in date sequence, which will be accomplished through the use of the unique identification key index.

The table in Figure 9.27 summarizes the results of reviewing the cost control system's database for access paths required by unique identification keys and Report and Inquiry sequences. The names assigned to the dBASE III PLUS index files are also shown. Required access paths should be included in the data dictionary as a note on the file's contents form.

Data File	Required Access Paths	Index File Name
ID	Account Number	IDIN1
BUDGET	Account Number+ Job Number+ Month	BUDIN1
EXPENSES	Date+ Account Number+ Job Number+ Source+ Description	EXPIN1
	Account Number+Date Job Number+Date	EXPIN2 EXPIN3

FIGURE 9.27

Cost control system access paths.

The Data Structure Diagram

The ID file and the BUDGET file are said to form a **one-to-many relationship** through the Account Number field because the ID file contains one record for each value of Account Number, and the BUDGET file contains many records for each value of Account Number. Similarly, the BUDGET file and the EXPENSES file also form a one-to-many relationship through the Account Number+Job Number+Month field combination. BUDGET contains one record for each Account Number+Job Number+ Month combination, and EXPENSES contains many records for each value of the combination.

While the contents forms are an effective tool for identifying and recording the contents of a system's database, they are not effective in expressing relationships among the data stores and files. For this purpose Pete uses a graphical model called a d**ata structure diagram**, Figure 9.28.

The diagram shows the structure of the cost control system database at a glance. The basic entity is the general ledger account represented by the file ID. For each general ledger account the master data store contains BUDGET records. The day-to-day transactions of the system are recorded in EXPENSES records, which are charged against the master data store BUDGET records.

The data structure diagram is developed by examining the unique identification key of each file. If many records exist in another file for each value of the key, then the files are connected on the data structure diagram with an arrow in the direction of the relationship. Data stores are indicated on the data structure diagram as **file superstructures**.

Summary

With the development of the data structure diagram, Figure 9.28, the design of the initial prototype's database is complete. Developing output data flows into contents forms and determining access paths and data structure diagram relationships are complicated activities. The rules and models presented in this chapter are designed to divide these activities into manageable tasks.

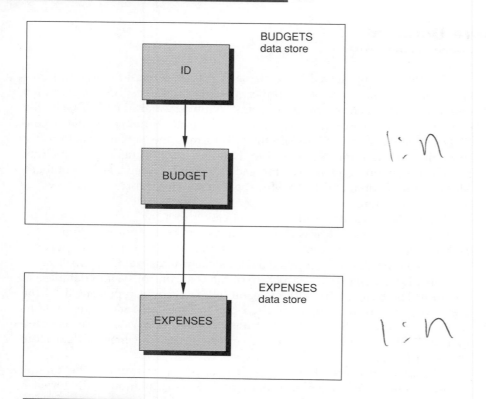

BUDGETS
data store

ID

BUDGET

$1:n$

EXPENSES
data store

EXPENSES

$1:n$

FIGURE 9.28

Data structure diagram for cost control system.

Designing the System Processes

The design of the software component of a transaction processing and/or reporting system, Pete and Betsy's next task, is easier than the design of the database. Using the DFD for a general transaction processing and/or reporting system, Figure 9.17, as a guide, the development team specifies the details of the system processes and organizes them into a model known as the system structure chart. The design effort begins with a review of the transaction maintenance process.

Step 1: Transaction Maintenance

In step one of database design Pete and Betsy identified BUDGETS as the master data store and EXPENSES as the transaction data store of the cost control system. So the process labeled MAINTAIN EXPENSES on the Design DFD, Figure 9.1, corresponds to the process labeled Transaction Maintenance on the DFD for a general transaction processing and/or reporting system, Figure 9.17.

A **transaction maintenance** module in a transaction processing and/or reporting system usually provides functions to Add records, Change records, Inquire about records, Delete records, and List records. The nature of the application, the programming environment, and the user/management group dictate the specific choices made by the builders. The choices, however, are all made within the context of the functions listed above.

For example, the experience of the user/management group may require functions such as change and delete to be combined into one menu choice, or the application may not allow transactions to be deleted once they have been entered.

Step 2: Master Maintenance

Master maintenance modules usually provide Add, Change, Inquire, Delete, and List functions also. Again the system builders make specific choices depending upon the nature of the application, the programming environment, and the user/management group.

Since Pete and Betsy identified the BUDGETS data store as the master of the cost control system, the process labeled MAINTAIN BUDGETS on the Design DFD, Figure 9.1, represents the process labeled Master Maintenance on the DFD for a general transaction processing and/or reporting system, Figure 9.17.

As a result of normalizing the database, the BUDGETS data store became two files, ID and BUDGET, so Pete and Betsy had to provide maintenance for both files under the MAINTAIN BUDGETS option of the cost control system main menu.

Step 3: Updating

Updating refers to the process of summarizing transaction data and recording the results in the master file. For example, an inventory system item master file record might contain a field called Quantity on Hand. Each time a sale of the item is made—i.e., a transaction occurs—the inventory system would decrease the Quantity on Hand master file field by the appropriate amount. If a new shipment of the item is received—i.e., another type of transaction occurs—then the system would increase the Quantity on Hand field by the appropriate amount.

Updating processes are included in transaction processing and/or reporting systems to simplify the reporting of master file **status information** such as inventory item quantity on hand. Without updating, the inventory system would have to compute the quantity on hand from the transaction data every time it was required.

The inclusion of an updating process in a transaction processing system is a tradeoff that the system builders must resolve. They must weigh the benefit of quick access to master file status information against the cost of programming the update and maintaining the accuracy of the status fields.

Transaction data volume also plays a role in the decision. The builders must weigh the benefit of the simple programming associated with eliminating the updating process against the cost of maintaining a complete file of transaction data for master file status computations.

In the cost control system, the BUDGET file record could contain a field called Actual Expenses, which would be updated by the system every time an expense transaction occurred. The benefits of such a design would be efficiency in the preparation of the Budget vs. Actual Reports and the possibility of deleting the transaction data after the update if the transaction detail is not needed for any other purpose. The costs of such a design would involve programming the update and maintaining the Actual Expenses status field accurately.

Pete and Betsy decided not to include an updating process in the design of the initial prototype. They reasoned that the transaction data had to be retained for the Inquiries and the Expense History Reports, and they could not foresee the need for more than a few Budget vs. Actual Reports per day. Even if they underestimated the report volume, the transaction volume of a few hundred per month would not create such a large EXPENSES file that reporting efficiency would suffer noticeably.

If the users experienced turnaround time problems during their use of the initial prototype, then the updating process could be added to a later version of the system. Unless transaction and reporting volumes demand the inclusion of an update process, system builders should choose ease of programming over efficiency considerations in the design of the *initial* prototype.

Step 4: Purging Records

Of course, system files cannot be permitted to grow indefinitely. Eventually, obsolete records must be deleted from the database. This process is known as **purging**, and it must be included in the design of every transaction processing and/or reporting system.

System builders must decide how often to purge and whether to automate the purge or present it as a menu option to be run at the user's discretion. Similar decisions must be made if the builders decide to include an update process in the design of the system.

In the cost control system the purging process is presented as option 6. PURGE JOBS on the cost control system main menu, Figure 2.1. The program deletes all records from EXPENSES and BUDGET corresponding to jobs that are more than two years old.

If a system employs an update process and maintains status fields in the master data store, then the purge process might also reinitialize these status fields. An example would be a year-to-date sales field that would be set back to zero as part of a year-end purging process.

In the data flow diagram for a general transaction processing and/or reporting system, Figure 9.17, purging is included in the update process.

Step 5: Reorganization

Maintaining an efficient database is a difficult task. As records are added, changed, and deleted from data files, dBASE III PLUS makes corresponding changes to the appropriate index files. Eventually, the complexity of the modified structures affects processing efficiency. When this happens, the database must be reorganized. In dBASE III PLUS, **reorganization** involves removing inactive data file records, if any, and rebuilding the index files.

As with updating and purging, system builders face the decisions of how often to reorganize and whether to automate the process or leave it to the user's discretion. In the cost control system the user can reorganize the ID, BUDGET, and EXPENSES files whenever he or she exits from a maintenance session in which one or more records were deleted. Since the files are small, the frequent reorganizations should not be cumbersome. If they are, then later versions of the system will present a modified reorganization strategy.

Process	Choice
Transaction maintenance	Add, Change, Inquire, Delete, List
Master maintenance	Add, Change, Inquire, Delete, List
Updating	YES/NO
	If YES, how often? Automatic or user initiated?
Purging	How often? Automatic or user initiated?
Reorganization	How often? Automatic or user initiated?

FIGURE 9.29

System process considerations.

As with purging, reorganization is included in the update process of the data flow diagram for a general transaction processing and/or reporting system, Figure 9.17.

The table in Figure 9.29 summarizes system process considerations.

The System Structure Chart

Before Pete and Betsy began programming the initial prototype, they organized the results of their work on system processes into a graphical model called a **system structure chart**. The chart represents the contents of system processes and the hierarchical relationships among the processes. System builders use the system structure chart to design the allocation of work to the processes and to determine the sequence of menus presented to the user.

For the initial prototype Pete and Betsy followed the processes of the data flow diagram for a general transaction processing and/or reporting system, Figure 9.17. They separated the reports process into two sections: Inquiries and Reports. User feedback on the initial prototype will determine if organization by the processes of Figure 9.17 is appropriate. The system structure chart is shown in Figure 9.30.

The system structure chart also serves as a programming aide. Wherever they could, Pete and Betsy wrote the name of the program next to the corresponding system structure chart box. Their naming convention used two letters and four numeric digits for programs beyond the main menu.

All of the programs called by the main menu were numbered with 0000. Programs called by a 0000 level program would be named with the same two letters as the 0000 level program and the digits 1000, 2000, 3000, etc. If the 1000 program calls programs, they would be numbered 1100, 1200, 1300, etc. If the 2000 program calls programs, they would be numbered 2100, 2200, 2300, etc. In this manner, the position of a program on the structure chart can be determined from its name.

Summary

This chapter began with techniques for the design of the data component of a transaction processing and/or reporting system. Contents forms for the Design DFD data stores were developed from the contents forms for the Design DFD output data flows.

The collection of contents forms represented a simple data dictionary. Contents forms data for the cost control system data stores will be entered into GENIFER's Data Dictionary module to begin the development of the software component. The contents forms data for the data flows will be stored in GENIFER's maintenance screen and report specifications.

Designing the dBASE III PLUS implementation of the data stores through normalization was discussed next. This was followed by an analysis of the system's Report and Inquiry sequences to determine the required access paths into the database. Database design concluded with a representation of the relationships among the database files in a data structure diagram.

Finally, system processes were determined through an analysis of the Design DFD, Figure 9.1, and the data flow diagram for a general transaction processing and/or reporting system, Figure 9.17. The processes were organized into a system structure chart according to the processes of Figure 9.17, and program modules were identified wherever possible.

Pete and Betsy had the initial prototype programs ready for the users in two working days. Data and index files were created through dBASE III PLUS commands, and a small collection of test data was compiled from the job cost ledger, time logs, and budget authorization sheets.

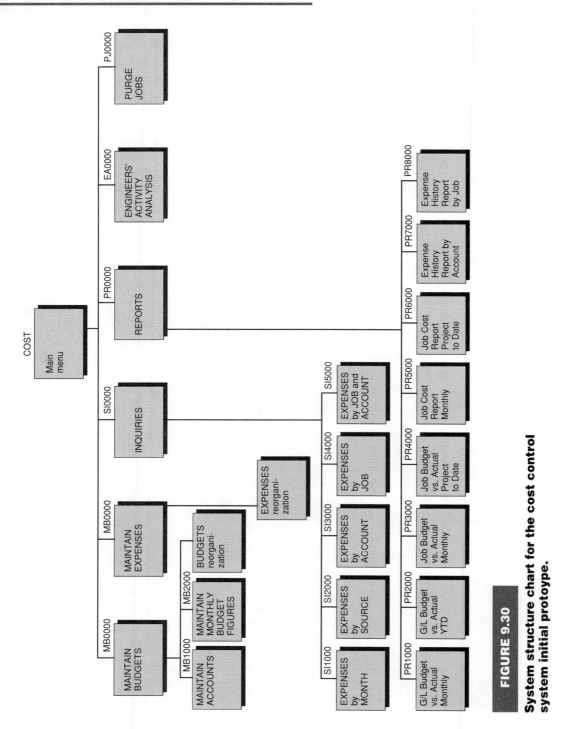

FIGURE 9.30

System structure chart for the cost control system initial protoype.

In the next chapter you will look inside the database and the programs of the initial prototype to appreciate the power of a fourth-generation/CASE software development environment. Upon completion of your study of the data and software, you will return to Horatio & Co. to watch the users experiment with the initial prototype and record their impressions.

THOUGHT QUESTIONS
In Your Opinion. . .

1. Betsy mentioned the difference between a management report and a report for accuracy control. What do you think is the difference? Why?

2. In a prototyping project the user fills the role of system designer. This being the case, do you think Pete was justified in developing a special format for the ENGINEERS' ACTIVITY ANALYSIS? Why or why not?

3. Which of the data store contents forms would you consider to be logical models of database components? Which would you consider to be physical models? Why? What about the data structure diagram shown in Figure 9.28?

4. What do you think are the advantages of normalizing the cost control system database? What do you think are the disadvantages?

5. What do you think are the tradeoffs involved in the decisions regarding automatic vs. user-initiated updating, purging, and reorganization?

References

1. Eckols, S. *How to Design and Develop Business Systems.* Fresno, CA: Mike Murach and Associates, Inc., 1983.

2. Kent, W. "A Simple Guide to Five Normal Forms in Relational Database Theory," *Communications of the ACM*, Volume 26, Number 2, February 1983, pp. 120–125.

3. Jenkins, A. M. "Prototyping: A Methodology for the Design and Development of Application Systems," *Spectrum*, Volume 2, Number 2, April 1985, pp. 1–8.

4. Jenkins, A. M. "Prototyping: A Methodology for the Design and Development of Application Systems—Part 2," *Spectrum*, Volume 2, Number 3, June 1985, pp. 1-4.

5. Weinberg, G. *Rethinking Systems Analysis and Design.* Boston: Little, Brown and Company, 1982.

HORATIO&CO.
EMPLOYEE CREDIT UNION

Data and Software Components

This chapter contained quite a bit of technical material. In this section you will apply the techniques of the chapter to the Horatio & Co. Credit Union System.

EXERCISES

1. Run every screen inquiry and every printed report in the credit union system. Complete a contents form for each of these output data flows.

2. Using the results of exercise 1 and the Design DFD you developed for the credit union system at the end of Chapter 7, complete a contents form for each of the credit union system's central data stores.

3. Design the database for the credit union system from the central data stores.

4. The actual database design for the credit union system is shown in Figures 9.1A through 9.4A. Determine the required access paths into the database through an analysis of the system's one-to-many relationships and the system's reporting and inquiry sequences.

5. Determine the credit union system's transaction maintenance, master maintenance, and updating processes through your experience as a system user.

6. Draw a preliminary system structure chart for the credit union system.

DATA STORE CONTENTS

Name: MASTER

Repetitions	*Component*	*Comments*
	Account Number	
	Name	
	Address	
	City_State_Zip	
	Opening Balance	As of 1st of month
	Interest Earned	Thru 1st of month
	Maintenance Fee Indicator	= YES if balance falls below $1000

underline = key component

	Index file name	*Index key*
	MASINI	Account Number

FIGURE 9.1A

Credit union system account master.

DATA STORE CONTENTS

Name: TRANS

Repetitions	*Component*	*Comments*
	Teller Code	
	Account Number	
	Date	
	Original/Reversal Ind.	O = original, R = reversal
	Type	D = deposit, W = withdrawal
		I = interest, M = maint. fee
	Amount	
	Cash/Check Indicator	C = cash, K = check

underline = key component
        (duplicate keys possible)

	Index file name	*Index key*
	TRIN1	Account Number + Date
	TRIN2	Teller Code + Date

FIGURE 9.2A

Credit union system transactions.

DATA STORE CONTENTS

Name: CODES

Repetitions	*Component*	*Comments*
	<u>Teller Code</u>	

underline = key component

	Index file name	*Index key*
	CODIN1	TELLER CODE

FIGURE 9.3A

Credit union system valid teller codes.

DATA STORE CONTENTS

Name: CONTROL

Repetitions	*Component*	*Comments*
	Current Month	Date of 1st of current month
	Cash Allowance	Tellers' starting cash balance

FIGURE 9.4A

Credit union system CONTROL data store.

CHAPTER 10

Inside the Initial Prototype

Objectives

The cost control system presented in Chapters 2 and 3 represented the initial prototype developed by Pete and Betsy. In those chapters you studied the system from the user's point of view. In this chapter you will study the system from the builder's point of view.

The cost control system was developed in an enhanced dBASE III PLUS environment. Although some environments provide greater capabilities, the cost control system environment is typical of modern software development. The knowledge gained from working with this environment on a microcomputer will be easily transferred to any other hardware/software configuration.

In an initial prototype the system builder tries to meet the user's basic requirements. Since Sam's primary concern is job cost reporting, this chapter concentrates on the inquiry and report programs and the database that supports them. In addition, this chapter considers the menu programs that organize the functions of the system structure chart, Figure 9.30, into a working system.

The specific objectives of this chapter are

1. To explore the cost control system software development environment, an example of a fourth-generation/CASE combination;

2. To observe the use of automated tools such as data dictionaries, report writers, and screen painters;

3. To examine the organization of the cost control system programs into a working system.

The Development
Environment: dBASE III PLUS
and GENIFER

dBASE III PLUS is a modern file management system. It provides the ability to create, maintain, and report from files of data records. It also provides the ability to connect files of related data for more sophisticated database processing and reporting. A review of the definition presented in Chapter 1 reveals that dBASE III PLUS qualifies as a simple fourth-generation environment.

dBASE III PLUS provides a programming language with **procedural** and **nonprocedural** commands. The procedural commands are similar to those found in such third-generation languages as COBOL, FORTRAN, and BASIC. The commands perform input/output, processing, and looping functions. The nonprocedural commands invoke sets of procedural commands through single words such as LIST (lists file records on screen), UPDATE (updates a master file from a transaction file), and TOTAL (summarizes detail records into a file of subtotal records).

GENIFER is a computer-aided software engineering (CASE) product that enhances the development of dBASE III PLUS applications. It provides processing and reporting capabilities not found in the dBASE III PLUS product, and it provides **automatic code generation**. Automatic code generation means that the system builder does not write programs. He or she describes system specifications, and the GENIFER code generation module automatically produces the necessary programs in the dBASE III PLUS programming language.

Obviously, GENIFER cuts system development time by eliminating program coding. In addition, the GENIFER database of specifications provides an effective source of system design documentation.

This **design database**, sometimes called a software engineering [2] or systems development [3] database, is the distinguishing characteristic of a CASE product [1]. The design database typically holds information about the data to be stored in the system, the business logic of the processes to be implemented, the physical layouts of the screens and reports, and other design information [1].

The GENIFER design database is comprised of a data dictionary and specifications for the system inquiries and reports, file maintenance, and menu programs. You will explore the inner workings of the initial prototype of the cost control system through a review of the GENIFER design database.

The Data Dictionary

Automated data dictionaries are used for recording, in a centralized location, all decisions related to the structure and implementation of every (system) file and record [2]. GENIFER is similar to other CASE tools in that the data dictionary forms the foundation upon which the design database is built. Betsy began the construction of the cost control system by entering a description of each system file into the data dictionary. Figure 10.1 shows a listing of the cost control system files in the GENIFER data dictionary. Like dBASE III PLUS, GENIFER refers to files as databases. Do not be confused by this.

Basically, a data dictionary system automates the maintenance of the information recorded in the contents forms introduced in Chapter 9. Figure 10.2 shows the GENIFER data dictionary entry for the ID file described in Figure 9.24. The information contained in the entry is typical of that stored by automated data dictionary systems. Field names are listed in alphabetical order. The numbers in the Key column indicate each field's place in the file key; a 0 in the Key column indicates that the field is not part of the file key. The Options column of the data dictionary entry is used to record data entry rules. Explanations of the picture codes used by dBASE III PLUS and GENIFER are provided in parentheses.

Figure 10.3 shows the GENIFER data dictionary entry for the BUDGET file described in Figure 9.25. Notice the BUDGET file key is made up of account number, job number, and month-ending date in that sequence.

GENIFER   Data Dictionary

No.	Database	Description
1	BUDGET	Monthly budget figures
2	CONTROL	Control data
3	EXPENSES	Expense transactions
4	ID	G/L account descriptions

FIGURE 10.1

Cost control system files in the GENIFER data dictionary.

Database ID   XX:XX XX/XX/XX   page 1

Description: G/L account descriptions

Field name	Type	Len	Dec	Key	Options	
ACCOUNT	Character	4	0	1	Picture: 9999	(four numeric characters only)
DESCRIPT	Character	30	0	0	Picture: @!	(alphanumeric characters: all alphabetic are uppercase)

FIGURE 10.2

The GENIFER data dictionary entry for ID.

Database BUDGET   XX:XX XX/XX/XX   page 1

Description: Monthly budget figures

Field name	Type	Len	Dec	Key	Options
ACCOUNT	Character	4	0	1	Picture: 9999 Valid values: ACCOUNT IN ID Error message: ACCOUNT NUMBER NOT FOUND IN ID
BUD_MNTH	Numeric	8	2	0	Picture: 99999.99
JOB	Character	4	0	2	Picture: @!
ME_DATE	Date	8	0	3	Picture: @D (date = XX/XX/XX)

FIGURE 10.3

The GENIFER data dictionary entry for BUDGET.

In addition to the data entry picture, the Options column for the AC-COUNT field specifies that any entry must be validated against the file ID. This means that an ID file record must exist for every value of ACCOUNT entered into a BUDGET file record. If an invalid ACCOUNT value is entered, the error message "ACCOUNT NUMBER NOT FOUND IN ID" is displayed.

Figure 10.4 shows the GENIFER data dictionary entry for EXPENSES. The EXPENSES key is made up of date, account number, job number, source, and description. ACCOUNT+JOB+month of DATE entries are validated against the BUDGET file. The AMOUNT and HOURS entries are validated against a numerical condition instead of a file.

The ability to store data entry and **data validation** rules with the definition of the file is a significant advantage of using a data dictionary system. Without the dictionary, data entry and validation rules must be written into each program, which usually leads to errors, omissions, and inconsistencies. Storing the rules in the data dictionary guarantees that every program invokes the same set of rules automatically.

The Report Writer

In Chapters 2 and 3, you saw that the output of the cost control system consisted of inquiries and reports. In a modern development environment one often produces inquiry and report programs with a software tool called a **report writer**. A report writer is a program that takes a description of an inquiry or report from the user and generates the necessary program from the description.

Both dBASE III PLUS and GENIFER provide report writers. The inquiries and reports of the cost control system were generated with the GENIFER report writer. It is comparable to many other commercially available products, so the knowledge gained from working with it will be easily transferred to any other environment.

The simplest report in the cost control system is the EXPENSES BY MONTH inquiry. Figure 10.5 shows an example that was run with a month-ending date of 03/31/89.

To use a report writer, the system builder divides the report into sections. In the EXPENSES BY MONTH inquiry the lines from the top of the report through the column headings are called the **report header**. The Grand total line is called the **report footer**, and the expense transactions lines are called the **detail**. To describe the sections to the report writer, the system builder creates a blank report layout with a word processor and

Database EXPENSES   XX:XX XX/XX/XX   page 1

Description: Expense transactions

Field name	Type	Len	Dec	Key	Options
ACCOUNT	C	4	0	2	Picture: 9999 Valid values: * Error message: **
AMOUNT	N	8	2	0	Picture: 99999.99 Valid range: AMOUNT >=0 Error message: AMOUNT MUST BE >= 0
DATE	D	8	0	1	Picture: @D Valid values: * Error message: **
DESCRIPTN	C	20	0	5	Picture: @!
HOURS	N	3	0	0	Picture: 999 Valid range: HOURS >= 0 Error message: HOURS MUST BE >= 0
JOB	C	4	0	3	Picture: @! Valid values: * Error message: **
SOURCE	C	17	0	4	Picture: @!

* = The combination of ACCOUNT+JOB+month of DATE in BUDGET.DBF
** = Error message: NO MATCHING BUDGET RECORD

FIGURE 10.4

The GENIFER data dictionary entry for EXPENSES.

then identifies each line as report header, detail, or report footer to the report writer.

Figure 10.6 shows the blank report layout for the EXPENSES BY MONTH inquiry.

```
DATE: XX/XX/XX                                               PAGE:    1
TIME:    XX:XX
                    EXPENSES BY MONTH FOR MARCH, 1989


SOURCE                ACCOUNT              JOB  DESCRIPTION        AMOUNT HOURS


ADAMS SUPPLY          4200 MATERIALS       B107 PROTOTYPE MATERIALS 4000.00     0
ETW LEASING           4300 EQUIPMENT       B107 CAD COMPUTER SYSTEM 4500.00     0
ASPEN ENGINEERING     4400 SUBCONTRACTORS  B107 CLEAN AIR DESIGN    5000.00     0
NAL-TECH              4200 MATERIALS       A141 TEST MATERIALS      2800.00     0
ADAMS SUPPLY          4200 MATERIALS       B762 FOUNDATION BLOCKS   2000.00     0
BOB JONES             4100 ENGINEERING     A141 DESIGN              4950.00    55
SARAH LUDWIG          4100 ENGINEERING     A141 STRUCTURAL TESTING  1200.00    30
SAM TILDEN            4100 ENGINEERING     B107 CONSULTATION        2400.00    40
BOB JONES             4100 ENGINEERING     B762 DESIGN               990.00    11
SARAH LUDWIG          4100 ENGINEERING     B762 DESIGN              2400.00    60


**   Grand total                                             30240.00   196
```

FIGURE 10.5

**EXPENSES BY MONTH for month-ending
date = 03/31/89.**

```
DATE: _____                            PAGE: ___
TIME: _____
                    EXPENSES BY MONTH FOR _____

SOURCE              ACCOUNT           JOB  DESCRIPTION        AMOUNT HOURS
_____ ____ _____ ____ _____ _____ ___
**   Grand total
                                                              _____ _____
```

FIGURE 10.6

**Blank report layout for EXPENSES BY
MONTH.**

Report items such as the column headings are constant; they remain the same no matter how many times the report is run. Report items such as the detail lines, however, are variable. They can change each time the report is run.

In creating the blank report layout for the report writer, the system builder types constant data where he or she wishes them to appear. The system builder usually indicates the location of variable data with a special symbol. The GENIFER report writer uses the underline character (_).

Report Data

Once the blank report layout is complete and the system builder has identified each line to the report writer, the report writer prompts the system builder to identify each variable field. The system builder has four choices: data dictionary data such as the detail line fields that come from the EXPENSES and ID files, system data such as the date and time, user request data such as the month-ending date for the inquiry, which is entered at the time the report is run, and computed data such as the grand totals.

With the specification of the variable data the report writer has enough information to write the report program. The advantage of using such a tool for prototyping is apparent. The system builder encourages the system user/designer to experiment with reports and inquiries because these programs can be produced in a matter of minutes. In many cases users learn to use report writers themselves for reports and inquiries that draw upon established databases.

The steps involved in using a modern report writer are summarized in Figure 10.7. The details may vary from one report writer to the next, but the concepts are essentially constant.

More Sophisticated Reports

Users of fixed-format reporting systems such as the cost control system rely upon the organizing powers of the computer to transform large volumes of data into meaningful information. An example of such a transformation is the **control-break report**.

1. Develop a blank report layout with a word processor.

2. Divide the blank report layout into sections and identify to the report writer the section to which each line of the layout belongs.

3. Identify the type and source of each variable data field.

4. Generate and use the report program.

FIGURE 10.7

Steps involved in using a report writer.

A control-break report is a columnar report that usually prints one line for every input record. The Expense History Report by Job, Figure 2.10, is an example of a control-break report. The input file for this report is EXPENSES. Notice that the lines of the report are grouped by JOB and within JOB by the month of the DATE. These are called the **control fields**.

Subtotaling represents the distinguishing characteristic of a control-break report. Examine Figure 2.10 and notice that it contains subtotals of $19,100 for Job A141, $27,800 for Job B107, and $7,600 for Job B762. These subtotals sum to $54,500, the total shown at the end of the report. The Expense History Report by Job shows a subtotal every time the control field, JOB, changes or breaks, hence the name **control-break report**.

A control-break report may have more than one level of subtotal. The Expense History Report by Job has two levels. Within Job A141, notice a subtotal for January of $10,100 and a subtotal for February of $9,000. These subtotals sum to $19,100, the subtotal for Job A141. The Expense History Report by Job produces a subtotal for every break in the control field month of the DATE. At this point you should verify that the subtotals for Jobs B107 and B762 equal the sum of the subtotals for January and February within each job.

Most elementary programming courses cover control-break logic in detail. With most report writers, however, control-break programming is not necessary. The system builder simply adds the descriptions of the control-break headers and footers to the report writer specifications. Both the dBASE III PLUS and GENIFER report writers handle control-break reports.

Figure 10.8 shows the specifications for the Expense History Report by Job, Figure 2.10, discussed above. Compare the specifications to the actual report.

```
DATE: _____                                                    PAGE: ___
TIME: _____
                          HORATIO & CO.
                       COST CONTROL SYSTEM
                   EXPENSE HISTORY REPORT BY JOB

DATE      SOURCE        ACCOUNT           DESCRIPTION        AMOUNT
                       Report Header
```

```
** Job  JOB
                  Break #1 Header - By Job
```

```
* MONTH

              Break #2 Header - By Month
```

```
DATE      SOURCE      ACCOUNT + NAME      DESCRIPTION        AMOUNT
                       Detail Line
```

```
* Subtotal for month                                        Subtotal
                  Break #2 Footer                           AMOUNT
```

```
** Subtotal for job                                         Subtotal
                  Break #1 Footer                           AMOUNT
```

```
*** Grand total                                             Grand total
                  Report Footer                             AMOUNT
```

FIGURE 10.8

**Report writer specifications for the
Expense History Report by Job.**

File Relations

The ACCOUNT columns of both the EXPENSES BY MONTH inquiry, Figure 10.5, and the Expense History Report by Job, Figure 2.10, illustrate an interesting phenomenon. The data dictionary entry for EXPENSES, Figure 10.4, does not show a field for Account Name, yet the name of the account charged for each expense appears in both the inquiry and the report. This is possible because of the relational database management capabilities provided by dBASE III PLUS.

Recall the normalization process discussed in Chapter 9. During that process Account Name was eliminated as a redundant nonkey component from all files except ID. The relational capabilities of dBASE III PLUS allow the system builder to relate the ID file to the EXPENSES file for reporting purposes. As a result the fields of the ID record corresponding to any EXPENSES record may be included in a report.

The correspondence between the files is made through the ACCOUNT number field, which both files contain. Whenever an EXPENSES record is encountered, dBASE III PLUS uses the value of the ACCOUNT field to retrieve the matching ID record and makes both records available to the report. The system builder specifies the desired relationship to the report writer as part of the report description process. Figure 10.9 shows the specification for the EXPENSES BY MONTH inquiry.

The **file relation** must be specified by the system builder because of the file orientation of dBASE III PLUS. Other database management systems provide this capability without a formal specification of a file relation. dBASE III PLUS also limits the system builder to one secondary file (database) for each primary file (database), while other database management systems are not so restrictive.

At this point you should be concentrating primarily on the inner workings of the cost control system. The details of implementing the system in dBASE III PLUS are secondary to the design concepts. Every hardware/ software environment presents strengths and weaknesses that the system builder must exploit and/or overcome.

File Maintenance

Options 1.   MAINTAIN BUDGETS and 2.   MAINTAIN EXPENSES of the cost control system main menu, Figure 2.1, represent the master maintenance and transaction maintenance modules of Figure 3.16 in the cost control system. The structure of any file maintenance module involves the

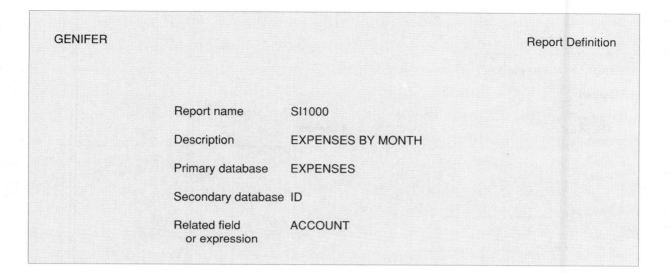

GENIFER   Report Definition

Report name   SI1000

Description   EXPENSES BY MONTH

Primary database   EXPENSES

Secondary database   ID

Related field   ACCOUNT
   or expression

FIGURE 10.9

**Report writer file specifications for
EXPENSES BY MONTH inquiry.**

familiar Add, Delete, Modify, Inquire, and List functions. To work
efficiently, the Change, Delete, and Inquire functions require a mechanism
to access the desired record(s). Usually, a **record retrieval by key value** and
a **browsing mechanism** are provided. The retrieval by key value is em-
ployed when the user has positive identification of the desired record, and
the browsing mechanism is used to search for records when positive
identification is not available.

Figure 10.10 shows the EXPENSES file maintenance screen
specifications for the cost control system. Like the report writer
specifications, the system builder types the headings and field descriptions
at the desired locations and marks the areas where the EXPENSES data
appears with underline characters. File maintenance specification modules
are sometimes called **screen painters**.

Once the maintenance screen is painted, the system prompts the system
builder for the type and source of the variable data that should appear at
each underline character location. The system builder supplies only the
field name; the row and column locations are determined by the system
from the painted maintenance screen. Field specifications for the EX-
PENSES file maintenance screen appear in Figure 10.11.

```
HORATIO & CO.
COST CONTROL SYSTEM
MAINTAIN EXPENSES

Date       _____


Account                              Account
Number     _____                    Name      _____


Job        _____


Source     _____


Description   _____


Amount     _____        Hours  ____
```

FIGURE 10.10

EXPENSES file maintenance screen.

Seq	Field name	Row	Col
1	DATE	5	6
2	ACCOUNT	9	9
3	DESCRIPT in ID	9	45
4	JOB	12	5
5	SOURCE	15	8
6	DESCRIPTN	18	13
7	AMOUNT	21	8
8	HOURS	21	43

FIGURE 10.11

EXPENSES file maintenance screen field specifications.

Data Validation

The data dictionary entry for EXPENSES, Figure 10.4, contained data validation rules for ACCOUNT+JOB+month of DATE, AMOUNT, and HOURS. The GENIFER program generator is able to access these dictionary rules and include the appropriate code in the file maintenance program. If an invalid entry is made, the appropriate error message is displayed, and the program executes a correction routine.

The system builder determines the appropriate file validation rules from the data structure diagram, Figure 9.28. In this figure the file ID is said to "own" the file BUDGET, and the file BUDGET is said to "own" the file EXPENSES.

File validation for a BUDGET record involves checking for a matching ID record through the value of ACCOUNT, the unique identification key of ID. In the same manner file validation for an EXPENSES record involves checking for a matching BUDGET record through the value of ACCOUNT+JOB+month of DATE, the unique identification key of BUDGET.

The general pattern of these rules involves checking the "owner" file of the file receiving the new record. The check looks for a record whose unique identification key value matches the values entered in the new record. The data structure diagram provides a useful guide to determining these file validation rules.

In the cost control system the validation of the ACCOUNT entry in the BUDGET file is absolute. A BUDGET record may not be entered for an ACCOUNT that does not exist in ID. The validation of the ACCOUNT+JOB+month of DATE entry in the EXPENSES file, however, is not absolute. The user is warned if no matching budget record exists, but he or she can override the warning and accept the expense. Betsy added this feature because unanticipated expenses that precede the necessary paperwork are common in construction projects.

In addition to data validation rule enforcement, Betsy included a **data verification** routine for Account Number in the cost control system. Data verification involves displaying descriptive information about data entry codes to ensure the accurate entry of the codes. The display is accomplished through the use of the DESCRIPT field from the ID file as a data entry aide. See field 3 in Figure 10.11.

Displaying the corresponding account DESCRIPTion when the user enters an ACCOUNT number for the EXPENSES record helps the user to enter transactions accurately. The routine requires a file relation between EXPENSES and ID with ACCOUNT number as the linking field.

File Indexes

The access paths developed in Chapter 9 are implemented in the cost control system through a dBASE III PLUS device known as the **file index**. A file index is a separate file associated with a data file. It serves the same purpose as the index at the back of a book. When the system needs to find a record matching a specified value, the system looks the record up in the index file first. The index file tells the system the number of the record it is seeking, and the system promptly sets the current record pointer to the appropriate value, thereby making the desired record available for processing. If a program traverses an index file sequentially, then the data file records appear in the sequence of the key of the index instead of the order of their entry into the data file.

The key of the primary index of the EXPENSES file of the cost control system is DATE+ACCOUNT+JOB+SOURCE+DESCRIPTN. This is the sequence used by the MAINTAIN EXPENSES and INQUIRIES modules. The cost control system also maintains EXPENSES indexes keyed by ACCOUNT and JOB for reports such as the Expense History Report by Account and the Job Cost Reports.

Whenever a data file is changed through file maintenance, the corresponding index files must be changed as well. Therefore file maintenance screen painters for systems that use index files must record the names of the index files associated with the data file in question. In GENIFER the identification of the index files is done at the time the file maintenance program code is generated.

A Word about Code Generators

It was mentioned earlier that GENIFER uses report writer and screen painter specifications to generate dBASE III PLUS programs automatically. The programs are physically identical to programs written by human beings. They may be modified by the human programmer and/or combined with hand-written programs to form a complete system.

Code generators work through skeleton files. Skeleton files contain programming modules for each task provided by the report writer or screen painter. The first step in code generation is the determination of which modules are needed to satisfy the specifications at hand. Once the modules are assembled, the code-generating software substitutes the particulars of the specification—such as file names, field names, and break fields—for the general parameters of the modules.

Several cost control system programs generated by GENIFER are included in the appendix at the end of this text. The portions of the programs typed in uppercase represent modifications made by Betsy Klein. Compare the programs to your own programming style, and keep in mind that these programs were actually written by a machine!

Tradeoffs: Code Generators vs. 4GLs

Code generators that produce programs in COBOL present an interesting alternative to fourth-generation languages. Three of the drawbacks to using a fourth-generation language are possible performance inefficiency, incompatibility with existing hardware and software, and training requirements. A COBOL code generator could be used to gain the development productivity of a fourth-generation environment without abandoning the existing hardware, software, or personnel. Wherever necessary, hand-written programs that optimize performance could be substituted for less efficient modules from the code generator.

While a COBOL code generator enhances programmer productivity, it does not provide a flexible tool for ad hoc reporting by users. If this capability is required, then a fourth-generation language must be included in the environment.

Menus

Menus represent the final component of the cost control system discussed in this chapter. Although they are discussed last in this chapter, system builders should implement the menu structure of a system immediately after the data dictionary. The system structure chart, Figure 9.30, serves as a guide.

In a fourth-generation/CASE environment, menu specification is straightforward. The system builder designs the menu layout with a word processor and then the system prompts him or her for the names of the programs to be run in response to the various user choices. Program code to check the validity of the user's entries is supplied automatically by the code generator.

Figure 10.12 shows the layout of the main menu of the cost control system. Since the user enters only his or her menu choice, the layout contains a single underline character after the prompt "Enter choice." Figure 10.13 shows the user choice specifications. Compare the figures to the cost control system structure chart, Figure 9.30.

Summary

This chapter presents a glimpse of the inner workings of the software component of the Horatio & Co. Cost Control System. The task is made easier by the organizing power of the GENIFER CASE tool. Accurate and

```
Menu: COST (MAIN MENU)                              XX:XX  XX/XX/XX  page 1
HORATIO & CO.
COST CONTROL SYSTEM
MAIN MENU



                    1.   MAINTAIN BUDGETS

                    2.   MAINTAIN EXPENSES

                    3.   INQUIRIES

                    4.   REPORTS

                    5.   ENGINEERS' ACTIVITY ANALYSIS

                    6.   PURGE JOBS


                    Q.   QUIT


                    Enter choice _
```

FIGURE 10.12

The cost control system main menu layout.

Entry	Program
1	MB0000
2	ME0000
3	SI0000
4	PR0000
5	EA0000
6	PJ0000

Menu exit key: Q   Default select key: Q   Erase before display: Y

FIGURE 10.13

The cost control system main menu choice specifications.

consistent maintenance of menu, screen, and report specifications is a significant advantage of CASE tools. More comprehensive and powerful CASE tools are discussed in Chapter 17.

The components of the cost control system software match the model of a transaction processing and/or reporting system presented in Figure 3.16. These systems are basically composed of file maintenance and inquiry and report modules accessed through a series of menus and integrated by the data dictionary.

Figure 10.14 presents the GENIFER main menu along with some notes about the definitions and customizer choices. Notice how the menu reflects the Figure 3.16 model. The prominence of the data dictionary is apparent. The menu, file maintenance, and report specifications are made through the Definitions function, and the programs are generated through choice 3. Program generators.

While the details might vary, all CASE tools that support the design and programming portions of the life cycle for transaction processing and/ or reporting systems follow the same pattern.

Once a development team determines that a transaction processing and/or reporting system provides a feasible solution to the problem and/ or opportunity at hand, their attention turns to making choices regarding system components to best meet the needs of the users. Although the Figure 3.16 model applies to all systems that provide transaction processing

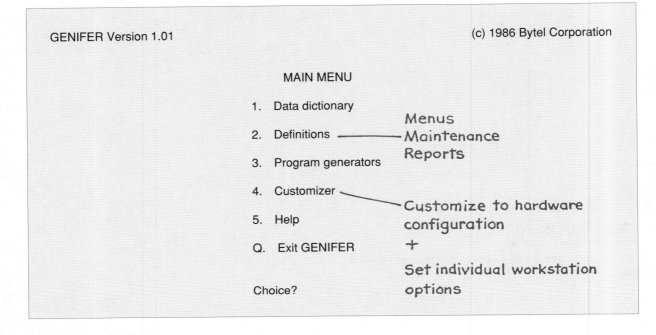

GENIFER Version 1.01   (c) 1986 Bytel Corporation

MAIN MENU

1. Data dictionary

2. Definitions ──────── *Menus*
                           Maintenance
3. Program generators   *Reports*

4. Customizer ──
                    ── *Customize to hardware*
5. Help   *configuration*

Q. Exit GENIFER   *+*

                        Set individual workstation
Choice?   *options*

FIGURE 10.14

The GENIFER main menu.

and reporting, choices regarding the details of each component are complex and far-reaching. The next chapter discusses how the prototyping design strategy supports this decision-making process.

THOUGHT QUESTIONS
In Your Opinion. . .

1. CASE tools that support the Systems Analysis Model Building Phase of the systems development life cycle exist. Based upon your experience with the Systems Analysis Phase of the cost control system project, what capabilities should these tools provide?

2. What is the role of the CASE design database and data dictionary in the Model Building Phase?

3. What kind of CASE tools might support the Evaluation of Alternatives Phase? the Problem and/or Opportunity Analysis Phase?

4. Do you think it is possible for CASE tools to generate system designs in the same way that GENIFER generates programs? Why or why not?

References

1. Gane, C. *Computer-Aided Software Engineering.* New York: Rapid System Development Inc., 1988.

2. Howden, W. "Contemporary Software Development Environments," *Communications of the ACM*, Volume 25, Number 5, May 1982, pp. 318–329.

3. Kozar, K. A. "A User Generated Information System: An Innovative Development Approach," *MIS Quarterly*, Volume 11, Number 2, June 1987, pp. 163–173.

HORATIO&CO.

Inside the Credit Union System

The credit union system is older than the cost control system. At the time the credit union system was developed, fourth-generation/CASE and microcomputer networking technologies were not as popular as they are today; thus the credit union software was programmed entirely by hand and designed to run on a timesharing network.

If the system were being developed today, Camille might consider running it on a microcomputer-based network. A decision in favor of the network would involve accepting more system responsibilities in return for lower operating costs. Regardless of how the credit union system was developed, the finished product can be discussed in light of the CASE technology presented in this chapter. The following exercises ask you to compose data dictionary entries and report writer specifications for various modules of the credit union system based upon your experience as user and developer.

In essence, the exercises ask you to "reverse engineer" the existing credit union system into the dBASE III PLUS/GENIFER environment. Some of the more sophisticated CASE products offer automated support for this process. See Chapter 17 for more information.

Exercises

1. Based upon your experience with the file maintenance modules of the credit union system, compose data dictionary entries for the files MASTER and TRANS.

2. Run the ACCOUNT INQUIRY option of the SCREEN INQUIRES several times. Once you understand what it does, formulate report writer specifications for it, using Figures 10.8 and 10.9 as guides. If you find it difficult or impossible to formulate the specifications, identify the features of the ACCOUNT INQUIRY that distinguish it from the cost control system reports.

3. Repeat exercise 2 for the CASH DRAWERS INQUIRY.

4. What file(s) do you think is (are) the source of the TOTAL ON_DEPOSIT REPORT?

5. What file(s) do you think is (are) the source of the YEAR END 1099s?

6. Develop report writer specifications for a program to print mailing labels for the credit union account holders. Identify the source file(s) for the report.

CHAPTER 11

Refining the Initial Prototype

Objectives

This chapter discusses the evaluation of the initial prototype and the development of subsequent versions of the system.

The specific objectives of this chapter are

1. To develop a framework for the evaluation of computer-based information systems;

2. To examine some of the advanced development techniques available in a fourth-generation/CASE environment;

3. To observe the interaction between the system user and the system builder in the prototyping process.

With this chapter Pete Willard begins to play a less important role in the cost control system project. Betsy Klein is the system builder. Her knowledge of dBASE III PLUS and GENIFER was an important factor in the decision to apply prototyping to the development of the software component of the system.

Pete's current responsibilities include training Betsy in the techniques of the prototyping process, helping with the development of the initial prototype, and consulting with Betsy on questions of design and implementation.

The User/Designer's Evaluation of the Initial Prototype

The plan for introducing the initial prototype of the cost control system to the users called for Betsy to work with Sam first and then with small groups of engineers. Whenever more than one user is involved in a prototyping project, it is customary to test the programs on a single representative user first and then introduce the system to the rest of the group [1].

Betsy used the data flow diagram of a general transaction processing and/or reporting system, Figure 3.16, to explain the organization of the cost control system's menus. She asked Sam to evaluate the components of the reports process, INQUIRIES and REPORTS, first. Evaluation of the maintenance processes would follow. She asked him to consider the system's organization, completeness, accuracy, and ease of use.

Betsy introduced the use of a "pop up" utility to Horatio & Co. several years ago. She suggested that Sam use the notepad option of the utility to record his thoughts while he was using the system. When they met as a team to review Sam's feedback, the notes provided the basis for their discussion. Sam's notes appear as Figure 11.1.

The System Builder's Refinement of the Initial Prototype

Prior to her **team review** meeting with Sam, Betsy examined Sam's notes and added the new inquiries and reports he requested to the cost control system. These included the INQUIRY BY DESCRIPTION, the Year-to-Date Job Budget vs. Actual Report, and the ENGINEERS' ACTIVITY ANALYSIS by Activity.

Adding a new report involved creating the report writer specifications, generating the report program, adding the report option to the appropriate menu screen, and regenerating the menu program. Figures 11.2 and 11.3 show the change to the INQUIRIES menu specifications as a result of the addition of the INQUIRY BY DESCRIPTION. Compare Figure 11.2 to the original INQUIRIES menu shown in Figure 2.11.

System Organization

Main menu easy to follow in light of DFD for transaction processing and/or reporting system.

Will inactive jobs be included in the Project-to-Date Job Cost Report and the Expense History Reports? If they are, is it possible to provide the option to eliminate them?

If I run an inquiry on the screen and decide to print it, I have to rerun the entire inquiry to get to the print option. Could we add a request to print at end of a screen inquiry to avoid rerunning inquiry?

Should be able to do more than one inquiry of a certain type before returning to INQUIRIES menu.

Completeness

Could use an option to print out all Job Budget vs. Actual Reports instead of entering single job numbers. Keep single job option and add all jobs option.

In ENGINEERS' ACTIVITY ANALYSIS could use an analysis by Activity showing who did what activity in addition to the analysis by Engineer showing what activities a person did. This would help with professional development plans.

Also, could use an INQUIRY BY DESCRIPTION to look at who sold us what. This would help with the reorganization of the vendor base.

Could use a Year-to-Date Job Budget vs. Actual Report in addition to the Monthly and the Project-to-Date Reports.

What about requests not available from the menus, like the average cost of engineering for office park construction projects?

Accuracy

How do I know INQUIRY BY SOURCE includes all records? What if an entry is misspelled when it is entered? Then it would not be included.

Asterisks on control-breaks and grand totals in inquiries and reports are inconsistent. If we follow the pattern of adding an asterisk for every new control break, then the grand total should have the least number of asterisks instead of the most.

Ease of Use

Would like the ability to page up and down through a long inquiry like a word processor.

Date of expense is missing from inquiries. Could we include it?

Could we have headings at the top of each screen or page in a long report or inquiry? Right now, we get headings on the first screen or page only.

FIGURE 11.1

Sam Tilden's notes on reports and inquiries.

```
Menu: SI0000 (INQUIRIES)                              XX:XX  XX/XX/XX  page 1
HORATIO & CO.
COST CONTROL SYSTEM
INQUIRIES MENU




              1.   EXPENSES BY MONTH

              2.   EXPENSES BY SOURCE

              3.   EXPENSES BY DESCRIPTION

              4.   EXPENSES BY ACCOUNT

              5.   EXPENSES BY JOB

              6.   EXPENSES BY JOB AND ACCOUNT


              Q.   QUIT



              Enter choice _
```

FIGURE 11.2

The updated INQUIRIES menu.

Betsy felt the assignment of the BY DESCRIPTION option to choice 3 was more logical than assigning it to choice 6. The GENIFER menu-generation module made the change simple, although Betsy realized with regret that her system structure chart, Figure 9.30, was obsolete. She modified the chart to reflect the additions to the system, but she was not optimistic about the prospect of keeping the chart current. At least the GENIFER menu documentation was always current. Betsy had heard about CASE tools that integrate DFDs and structure charts with the data dictionary and system specifications, and now she realized how valuable these tools really were.

Entry	Program
1	SI1000
2	SI2000
3	SI6000 (New INQUIRY inserted at choice 3)
4	SI3000
5	SI4000
6	SI5000

Menu exit key: Q   Default select key: Q   Erase before display: Y

FIGURE 11.3

The updated INQUIRIES menu choice specifications.

The Year-to-Date Job Budget vs. Actual Report was added to the RE-PORTS menu in the same way. The new Engineers' Analysis by Activity required the creation of an ENGINEERS' ACTIVITY ANALYSIS menu showing two analysis choices: BY ENGINEER and BY ACTIVITY. Betsy generated the appropriate menu and report programs and updated her system structure chart as well.

The Team Review

Betsy began her meeting with Sam by showing him the new inquiries and reports on the computer. He was pleased with them and impressed with the speed with which she had developed them.

After the demonstration of the new inquiries and reports, Betsy turned her attention to Sam's comments that concerned the output component as a whole. These were the comments regarding headings on every page of long inquiries and reports, print requests at the end of a screen inquiry, adding the DATE field to all inquiries, asterisks on the Grand total line, and paging up and down through a long screen inquiry or report.

Betsy explained her plan to Sam. She wanted to modify the INQUIRY BY ACCOUNT program to address Sam's comments as a test case. Sam could experiment with the modified program and provide feedback to resolve the issues. Once Sam was satisfied with the INQUIRY BY ACCOUNT,

Betsy would incorporate his decisions into the remaining inquiries and reports.

Sam agreed with the plan, and Betsy brought up the blank INQUIRY BY ACCOUNT layout on the screen. Betsy realized Sam's comment about the inconsistent use of the asterisks was correct, so she deleted the asterisks from the Grand total line of the layout.

Next Betsy changed the identification of every line of the report header section to page header. She explained that report headers are displayed once at the beginning of an inquiry or report, while page headers are displayed at the top of every page. Sam's request called for page headers; Betsy had originally specified report headers.

With the completion of these two specifications, Betsy turned to the detail line. The line contained the SOURCE, ACCOUNT NUMBER+NAME, JOB, DESCRIPTION, AMOUNT, and HOURS fields. From the data dictionary entries for EXPENSES and ID, Figures 10.4 and 10.2, Betsy determined the length of the data to be 17+4+30+4+20+8+3=86 characters. An additional 6 characters accounted for the spaces between the fields, bringing the total to 92. To fit the original detail line on an 80 character screen, Betsy had originally cut 14 characters from the 30 character Account Name display. Adding a DATE field now would require an additional 9 characters. (Why?)

The Account, Job, Amount, and Hours displays could not be cut, and Sam thought the Source and Description displays were short enough and did not want them cut. This eliminated the possibility of inserting the DATE field on the existing line.

Betsy offered the possibility of a two-line detail section. Sam was intrigued by this possibility, so Betsy set it up on the layout and generated the program. Since she was running this session from a test diskette, she was not afraid to make radical changes to the system specifications. Changes to the working copy of the system would be made only after Sam reviewed them.

Unfortunately, Sam did not like the two-line detail section, so he gave up on the idea of a DATE field in the inquiries. Betsy returned the blank layout to the one-line configuration.

Betsy showed Sam an idea she had to send inquiries to a file instead of the screen and then access the file through a word processor to provide the ability to page up and down through the inquiry. Sam felt the arrangement was too cumbersome, so Betsy dropped the idea.

Betsy was able to modify the GENIFER program to provide a print request after a screen inquiry, but she asked Sam to give up his request to make more than one inquiry of a given type before returning to the INQUIRIES menu. This request would involve changing the fundamental logic of the GENIFER programs and would defeat the purpose of using the code generator. Sam agreed that the benefits of such a change were not worth the effort.

This request was the last on Betsy's agenda for this team review session. She told Sam she would work on the other items and meet with him again in a few days. Before she left, Betsy gave Sam a preview of the ad hoc reporting capabilities available through the dBASE III PLUS command level. She used Sam's example of calculating the average cost of engineering for office park construction projects.

After reading the next section, try the example yourself. Load dBASE III PLUS as usual, but do not activate the cost control system by typing DO COST. Type the commands shown in Figure 11.4 instead.

The dBASE III PLUS Command Level

The initial prototype of the cost control system is made up of programs written in the dBASE III PLUS programming language. Programming is one mode of dBASE III PLUS operation. dBASE III PLUS may also be used as an interactive database management system. This means that commands entered from the keyboard are executed immediately and the results displayed on the screen or printer. This mode of operation is known as **command level processing**.

Figure 11.4 shows a dBASE III PLUS command level session in which Betsy determined the average cost of engineering for office park construction projects and then made some additional calculations. Betsy knew that account number 4100 represented Engineering charges. Sam informed her that the office park construction projects in the database were Jobs A141 and B762.

While the commands shown in the figure are dBASE III PLUS commands, the concepts of command level processing presented in the example apply to all fourth-generation environments. The lines that begin with a . and contain uppercase characters are those typed by the user. The lowercase lines represent the responses displayed by dBASE III PLUS.

The first command selects the file EXPENSES under the control of index EXPIN3 for use. The key of index EXPIN3 is JOB+DATE. Subsequent commands will operate on the file EXPENSES until another USE command is issued.

To find the average engineering cost per job, Betsy needs to calculate the total engineering cost for each job. The second command does this; it writes the results in a dBASE III PLUS file called TEMP.

The TOTAL ON command requires the data file to be under the control of an index whose key matches the field specified as the TOTAL ON control field. In this example the TOTAL ON control field is JOB, which matches the key of index EXPIN3.

Subsequent commands refer to TEMP, so TEMP is put into use in the third command. The fourth command calculates the average value of the AMOUNT for the TEMP records which represent office park construction projects. The fifth command repeats the calculation for the HOURS field. The ? represents the dBASE III PLUS command for calculate. The result is the average hourly cost of engineering for office park construction projects.

Prototyping Concepts

The preceding section shows the interaction between the user/designer and the system builder in the prototyping process [1]. The user is the designer of the "front end" of the system. He or she controls the design process.

```
. USE EXPENSES INDEX EXPIN3

. TOTAL ON JOB TO TEMP FOR ACCOUNT='4100'
TEMP.dbf already exists, overwrite it? (Y/N) Yes
     15 Record(s) totaled
      3 Records generated

. USE TEMP

. AVERAGE AMOUNT FOR JOB='A141' .OR. JOB='B762'
      2 records averaged
  amount
15120.00

. AVERAGE HOURS FOR JOB='A141' .OR. JOB='B762'
      2 records averaged
  hours
    243

. ?15120/243
     62.22
```

FIGURE 11.4

A dBASE III PLUS command level session.

When Sam decided he did not like the two-line detail section, Betsy dropped the idea and moved on to something else.

The system builder assumes the more technical role of constructing a working system that satisfies the user/designer's requirements. The simple two-person development team provides the opportunity for effective team reviews of project work to date.

The first review of the initial prototype examines the organization, completeness, accuracy, and ease of use of the system's output component. The user supplies feedback on the functions the system provides and the format in which it provides them. This function/format combination is sometimes called the **user interface**.

In developing a user interface, system designers trade off flexibility and speed with security and simplicity. Naturally, cost is also a consideration. Inexperienced users who use a system infrequently want the security of a menu-driven system with simple menus to guarantee the successful completion of their work. Experienced users want the speed and flexibility of issuing their own commands directly to the system instead of wading through a series of menus.

The cost control system development team recognized the need for both types of user interface early in the process. The menus provide easy access to a variety of frequently used reports and inquiries (simplicity). Ad hoc reports, such as the average cost of engineering for office park construction projects, require use of the dBASE III PLUS command level (flexibility). By his comments, Sam Tilden seems to be a potential user of both interfaces.

The preceding section also shows the responsiveness of the prototyping process to user needs. In many cases the system builder can exploit the fourth-generation/CASE technology to satisfy user requests immediately.

Other requests, such as the elimination of inactive jobs from the Project-to-Date Job Cost Reports, cannot be delivered immediately, but certainly within a day or two. These requests are discussed in the next section.

Finally, some items, such as Sam's concern about spelling errors in data entry, are beyond the scope of the current review and must wait until later versions of the system. These concerns are discussed at the end of the chapter.

Continuing to Refine the Initial Prototype

Back in her office Betsy reviewed the notes from her meeting with Sam. She incorporated the changes in the INQUIRY BY ACCOUNT print option, Grand total, and page headings into the other inquiry and report layouts and generated a new set of inquiry and report programs. She also checked her system structure chart against the GENIFER documentation for consistency.

When Betsy added these changes to the new inquiries and reports she had developed before the meeting and the requests that Sam subsequently decided to drop, she realized only three items from Sam's original list, Figure 11.1, remained unresolved. These were the option to eliminate inactive jobs from the Project-to-Date Job Cost Report and the Expense History Reports, the option to print all Job Budget vs. Actual Reports at once, and the concern for the effect of variations in the spelling of Source and Description entries in EXPENSES.

Inactive Jobs

Sam's use of the initial prototype brought the idea of an inactive job into focus. Prior to this time inactive jobs were considered in relation to the PURGE JOBS option of the cost control system main menu, Figure 2.1. Since that option would have no effect on the system for the next two years, it did not receive much attention.

Now the inactive job was seen as a threat to the effectiveness of the cost control system. If the Project-to-Date Job Cost Report was cluttered with inactive jobs, then its usefulness was diminished. If Sam and the project managers had to remember or guess which jobs were active and which were inactive, then the cost control system could actually do more harm than good.

Betsy knew she had to resolve this situation. Her instincts told her an active/inactive indicator for each job should be stored somewhere in the database. She thought back to the ideas she had discussed with Pete Willard in relation to the general model of a transaction processing and/or reporting system, Figure 3.16.

In the model the MASTER data store contains background and/or status information about system entities, and the TRANSACTION data store captures the day-to-day activity of the system. In the cost control system the Budgets data store represents background information about general ledger accounts and jobs, while the EXPENSES data store represents day-to-day activity against the accounts and jobs.

The active/inactive indicator seemed to be a piece of status information about a job. Betsy reviewed her cost control system data structure diagram, Figure 9.28. She decided to add a file called JOB to the BUDGETS data store. JOB is the analog of ID; it holds background and status information about jobs in the same way that ID holds this information about G/L accounts. Figure 11.5 shows the updated data structure diagram.

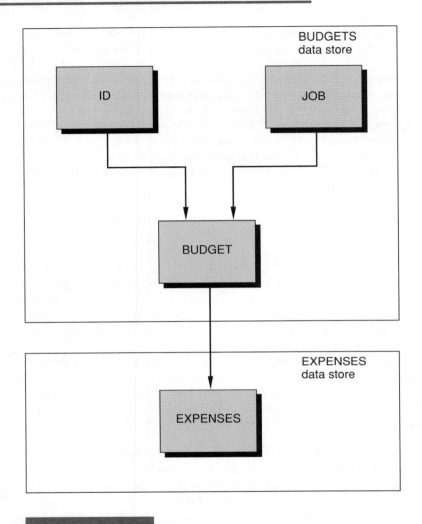

FIGURE 11.5

**The updated data structure diagram for the
cost control system.**

Betsy entered the layout of JOB into the data dictionary. Initially, she
entered two fields: JOB NUMBER and the active/inactive indicator.

The addition of a new file to the system required changes to the file
maintenance procedures as well. Since JOB was included in the BUDGETS
master data store, Betsy added an option called MAINTAIN JOBS to the
MAINTAIN BUDGETS file maintenance menu. She developed a file maintenance screen with the screen painter and called the program MB3000.

Figures 11.6 and 11.7 show the changes to the MAINTAIN BUDGETS menu specifications.

With the completion of the specifications, Betsy generated the new file maintenance program and regenerated the updated menu program. As a test she entered JOB records for A141, B107, and B762 and set the active/inactive indicator to active for all three. Since Sam was assigned responsibility for the BUDGETS data store, he will be responsible for the accurate maintenance of the indicators. Betsy examined the file JOB from the dBASE III PLUS command level to determine the accuracy of her work. She also noted the changes to the cost control system on her system structure chart.

Finally, Betsy modified the specifications for the Project-to-Date Job Cost Report and the Expense History Reports to prompt the user for his or her choice regarding inactive jobs. According to the choice, inactive jobs are included or not included in the reports.

```
Menu: MB0000 (BUDGETS DATA STORE MAINTENANCE)          XX:XX  XX/XX/XX  page 1
HORATIO & CO.
COST CONTROL SYSTEM
MAINTAIN BUDGETS MENU




                  1.   MAINTAIN ACCOUNTS

                  2.   MAINTAIN JOBS

                  3.   MAINTAIN MONTHLY BUDGET FIGURES


                  Q.   QUIT




                  Enter choice _
```

FIGURE 11.6

The updated MAINTAIN BUDGETS menu.

Entry	Program
1	MB1000
2	MB3000 (NEW PROGRAM INSERTED AT CHOICE 2)
3	MB2000

Menu exit key: Q   Default select key: Q   Erase before display: Y

FIGURE 11.7

**The updated MAINTAIN BUDGETS menu
choice specifications.**

The changes to the working copy of the cost control system had taken the entire afternoon. Betsy decided to quit at this point and attack the remaining two points in the morning. She was very satisfied with the day's work.

Job Budget vs. Actual Reports

Given the design of the cost control system database, the Budget vs. Actual Reports could not be produced by the GENIFER report writer. (Why?) Betsy added a series of dBASE III PLUS TOTAL commands to the report programs to manipulate the system data into a temporary holding file from which the reports could be generated. The technique is similar to manipulations shown in Figure 11.4. In this case the TOTAL ON control field was ACCOUNT.

For the Job Budget vs. Actual Reports, Betsy used the Job Number entered by the user in a FOR clause of the TOTAL commands to include only records that matched the selected job in the calculations. Again, see Figure 11.4 for an example of the FOR clause.

To produce a Job Budget vs. Actual Report for all jobs automatically, Betsy changed the TOTAL ON control field to the JOB+ACCOUNT combination and added a control break by JOB to the original Job Budget vs.

Actual report writer specifications. (Why?) She generated the reports as separate programs and entered them on the REPORTS menu as ALL JOBS BUDGET vs. ACTUAL REPORT MONTHLY and ALL JOBS BUDGET VS. ACTUAL REPORT PROJECT-TO-DATE.

The use of the JOB+ACCOUNT combination in the TOTAL ON command required the addition of a JOB+ACCOUNT field to both the BUDGET and EXPENSES files. The reason for including the JOB+ACCOUNT field in the records is that the TOTAL ON command requires a data file field as the TOTAL ON control. The control field cannot be a combination of fields; all of the control data must reside in a single field. This is a common occurrence in fourth-generation systems.

Betsy modified the data dictionary entries and the file maintenance programs accordingly. The user did not have to enter the combined field; it was computed automatically from the entries in the individual JOB and ACCOUNT fields.

The Second Prototype

The last remaining item from Sam's list of comments, Figure 11.1, was the concern over spelling errors at data entry time. Since this concerned the file maintenance processes, which had not been reviewed yet, Betsy decided to deal with this comment in the next round of user feedback. The development of the second prototype of the cost control system was therefore complete.

The development of the software component from beginning through the second prototype took 10 working days. The rapid turnaround of the prototyping process depends upon the fourth-generation environment and the simple one-to-one interaction between the system user and the system builder. In the next section Betsy turns her attention to the more difficult problem of communicating with a group of system users.

Working with Secondary Users

When she added the new Job Budget vs. Actual Reports, Betsy fulfilled every one of Sam's requests except the one dealing with spelling errors in the data entry process. Sam reviewed the inquiries and reports of the second prototype and found them to be satisfactory. In that time Betsy received and installed the network version of dBASE III PLUS. She set up a user profile for the engineers that allowed read-only access to the ID file, the JOB file, and the data of the EXPENSES file except for the SOURCE field. She did this to protect the accuracy of the data and the confidentiality of the salary information.

The engineers needed to use the cost control system Inquiries function only. Since they did not have access to the SOURCE field, they could not do an INQUIRY BY SOURCE. Betsy created a modified version of the IN-QUIRIES menu that did not include the INQUIRY BY SOURCE. She called the program INQUIRE. It served as the main menu of the engineers' version of the system. To run their version of the system, the engineers were instructed to type DO INQUIRE.

The modular design of the cost control system programs and the nor-malized design of the database allow selective access to system functions and data. System builders can produce customized interfaces easily through the manipulation of the top-level menu programs of the system structure chart. This flexibility is one of the rewards of following the de-manding, and at times tedious, procedures of the Analysis and Design Phases of the development process.

Betsy introduced the system to the engineers in groups of four. After a short demonstration she let members of the group pose job cost questions, which the group analyzed together. The engineers took turns entering specifications into the system and verifying the results [3]. At the comple-tion of the team review, Betsy suggested the use of the "pop-up" notepad for recording individual impressions.

The engineers worked with the second prototype about two weeks. A review of the comments revealed no changes to the menu-driven system. They did not miss the INQUIRY BY SOURCE. All of the unsatisfactory comments referred to analytical situations best handled at the dBASE III PLUS command level.

Betsy scheduled the command level training to begin in two weeks. She left a number of copies of a dBASE III PLUS book with a good tutorial near the engineers' workstations. She wanted to take advantage of the engineers' curiosity because she believed in learning through discovery as well as through classroom instruction.

Betsy planned to use the two-week period before the start of the com-mand level training to work with Sam on the evaluation of the maintenance processes of the system. These processes correspond to those shown at the top of Figure 3.16, the data flow diagram for a general transaction process-ing and/or reporting system.

The User/Designer's Evaluation of the Second Prototype

While the engineers were reviewing the output component of the second prototype, Sam Tilden was experimenting with the maintenance procedures for the BUDGETS data store. Betsy advised the same four criteria: system organization, completeness, accuracy, and ease of use. Sam's impressions, recorded on the "pop-up" notepad, appear in Figure 11.8.

The System Builder's Refinement of the Second Prototype

Betsy was able to respond to Sam's comments quickly. She went back to the data dictionary entry for the BUDGET file. Now that the JOB file had been

---

System Organization

   Organization is simple and clear.

Completeness

   Functions seem to be complete.

Accuracy

   Why does the system let me enter invalid job numbers?

Ease of Use

   Blank records appear mysteriously in file.
   What is the sequence of the records?

FIGURE 11.8

Sam Tilden's notes on maintenance procedures.

added to the system, she could enter a validation condition for the JOB field of BUDGET which was similar to the ACCOUNT field condition: every job number entered into a BUDGET record must match an existing JOB file record. She regenerated the BUDGET file maintenance program and showed Sam how it worked.

While she was at the computer, she explained the reason for the blank records in the BUDGET file. Blank records occur when the user does not exit from the Add function properly. The file maintenance programs are designed to accept new records until the user indicates the end of the session by hitting Esc on a blank record. If the user hits some other key, then it is possible for the blank record to be added to the file.

Betsy changed the specification of the maintenance program to accept one new record and return to the maintenance option menu screen automatically. This reduced the chance of entering a blank record, but slowed the data entry process. Sam usually entered new records in batches. Under the new arrangement he had to activate the Add function for each new record. He asked Betsy to return the maintenance program to its original design, and she did.

Finally, Betsy explained that the sequence of the records in all of the maintenance programs was the unique identification key, the string of fields whose values uniquely identify a record. For ID, the identification key is ACCOUNT. For JOB, it is JOB, and for BUDGET, it is ACCOUNT+ JOB+ME_DATE.

The Third Prototype

While Sam was experimenting with the maintenance procedures for the BUDGETS data store, Betsy was doing the same with the EXPENSES data store. Betsy realized that the SOURCE and DESCRIPTN fields needed to be validated. This condition led to Sam Tilden's comments about spelling errors in his review of the initial prototype.

Since the cost control system design called for the maintenance of expense transactions through the central accounting system, Betsy decided not to implement SOURCE and DESCRIPTN validation in the microcomputer-based program. She would include the validation conditions in her specifications to the programmer who was modifying the central accounting system.

With that decision the development of the third prototype of the cost control system was complete.

Summary

This chapter discusses the evaluation of computer-based information systems based upon four criteria: system organization, completeness, accuracy, and ease of use.

Several points regarding the interaction of the system user and the system builder after the development of the initial prototype are worth noting.

1. The user controls the design process.
2. The modular design of the program structure and the normalization of the database provide the flexibility necessary for rapid turnaround on user feedback.
3. Systems design is a continual series of tradeoffs.
4. The user and the builder must develop a relationship based upon cooperation and trust. The best way to do this is to work together through all phases of the project.

THOUGHT QUESTIONS
In Your Opinion. . .

1. When Betsy developed the INQUIRY BY DESCRIPTION, she placed the menu option at choice number 3. Do you think this is the proper place for it? Does the organization of menu options really matter?
2. Based upon your answer to the questions above, where would you place the new ALL JOBS BUDGET vs. ACTUAL REPORTS on the REPORTS menu. Why?
3. Betsy Klein modified many of the programs generated by GENIFER for the cost control system. If a program is regenerated during the prototyping process, then Betsy's modifications must be redone. What would be the most convenient and accurate way to keep track of modifications to generated programs?

References

1. Jenkins, A. M. "Prototyping: A Methodology for the Design and Development of Application Systems," *Spectrum*, Volume 2, Number 2, April 1985, pp. 1–8.

2. Jenkins, A. M. "Prototyping: A Methodology for the Design and Development of Application Systems—Part 2," *Spectrum*, Volume 2, Number 3, June 1985, pp. 1–4.

3. Kraushaar, J., and L. Shirland. "A Prototyping Method for Applications Development by End Users and Information System Specialists," *MIS Quarterly*, Volume 9, Number 3, September 1985, pp. 189–197.

4. Weinberg, G. *Rethinking Systems Analysis and Design*. Boston: Little, Brown and Company, 1985.

HORATIO&CO.
EMPLOYEE CREDIT UNION

Evaluating the Credit Union System

The following exercises ask you to evaluate the credit union software according to system organization, completeness, accuracy, and ease of use. The four-component model is a valid evaluation tool regardless of the software development methodology. Of course, in a prewritten software package project, this analysis would be made in the Evaluation of Alternatives phase of the life cycle, before the package was purchased.

These evaluation exercises discuss the similarities and differences between the cost control and credit union systems. Although the systems have much in common, features such as system reports and maintenance procedures are specific to the purpose of each application. These observations point up the need to keep the original objectives and tactics in mind when designing and/or evaluating systems.

Exercises

1   Evaluate the output component of the credit union system. Use the four criteria discussed in the chapter: system organization, completeness, accuracy, and ease of use.

2. Redo exercise 5, Chapter 7, in light of your experiences in this chapter. Discuss the tradeoffs involved in deciding to purchase a package instead of building the software component of a system.

3. Discuss the tradeoffs involved in deciding to use a fourth-generation language and prototyping as opposed to a third-generation language and the detailed design method.

4. Evaluate the transaction maintenance component of the credit union system.

5. Evaluate the master maintenance component of the credit union system.

6. Compare the user interface of the maintenance processes of the cost control system and the credit union system. Name two similarities and two differences.

PART FOUR

Completing the Systems Development Life Cycle

Part Four of this text contains three chapters. Chapter 12 presents an analysis of the technical issues associated with the day-to-day operation of computer-based information systems. Chapter 13 considers the integration of the completed system into the existing infrastructure of the organization, and Chapter 14 reviews the development of the Horatio systems to develop a general framework of systems design tradeoffs.

Chapter 14 marks the beginning of our movement away from the Horatio & Co. systems toward more general considerations. Part Five, the final part of this text, completes the shift by synthesizing the Horatio & Co. experiences into a general discussion of technical and organizational issues.

4

CHAPTER 12

Systems Implementation

Objectives

This chapter discusses the steps Betsy Klein took to make the prototype of the cost control system operational. At this point the user requirements are satisfied. Betsy must now consider the technical issues associated with the day-to-day operation of a computer-based information system.

The specific objectives of this chapter are

1. to determine the interface of the cost control system with the central accounting system;

2. to accomplish the conversion of manual job cost data to the automated system;

3. to develop realistic data entry controls;

4. to evaluate the processing efficiency of the cost control system with the working database in place.

The Fourth Prototype

The review of the initial prototype of the cost control system concentrated on the organization, completeness, accuracy, and ease of use of the inquiries and reports. The feedback was incorporated into the design of the second prototype.

The review of the second prototype concentrated on the maintenance processes. Sam Tilden was responsible for the maintenance of the BUDGETS data store, and Betsy Klein was responsible for the maintenance of the EXPENSES data store.

Sam's feedback on BUDGETS maintenance was incorporated into the design of the third prototype, and he was satisfied with the improvements. Because EXPENSES maintenance involved the central accounting system, Betsy chose to postpone the implementation of some of her feedback until the development of the new accounting system programs, which would be incorporated into the fourth prototype of the cost control system.

EXPENSES data presented a special problem because this data overlapped the data of the Accounts Payable and Payroll modules of the minicomputer-based central accounting system. The cost control system could not run independently of the accounting system for long before differences in the common data began to appear.

To avoid this problem, the development team suggested the modification of the accounting system maintenance modules to accept the job cost data along with the regular accounts payable and payroll data. Once entered and checked, EXPENSES data store records could be transferred electronically from the accounting system minicomputer to the cost control system microcomputer network. The design of the interface between the two systems is discussed in the following sections.

The Central Accounts Payable System

The Horatio & Co. accounting system is composed of an integrated set of modules for General Ledger, Accounts Receivable, Accounts Payable, and Payroll processing. Cost control system expense data originates in the Accounts Payable and Payroll modules.

The Accounts Payable (A/P) module is a transaction processing system, but it is more complicated than either the cost control system or the credit union system. A level 0 data flow diagram for the Accounts Payable module is shown in Figure 12.1.

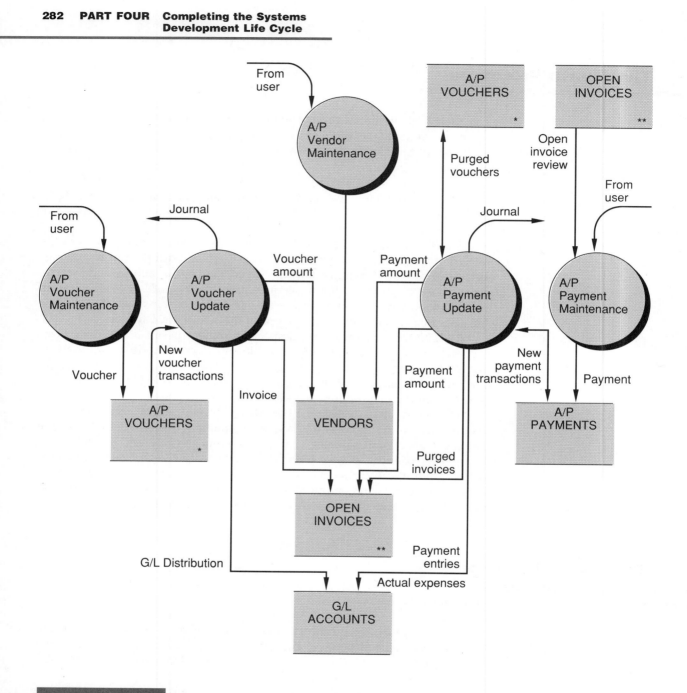

FIGURE 12.1

Level 0 DFD for Accounts Payable module.

Accounts Payable uses three master data stores: VENDORS, OPEN IN-VOICES, and G/L ACCOUNTS. A/P Vendor Maintenance provides the typical maintenance functions—Add, Delete, Modify, Inquire, and List—for the VENDORS data store. No manual maintenance is allowed on the OPEN INVOICES data store; changes to INVOICES must be achieved through transactions. G/L ACCOUNTS maintenance is performed through the General Ledger module.

The system handles two types of accounts payable transactions: vouchers and payments. A voucher is a statement by the responsible manager that a vendor invoice is legitimate and, therefore, eligible for entry into the accounts payable system. This communication is achieved through the payment authorization voucher, Figure 5.9, from the manual cost control system. After Sam Tilden completes a voucher, Betsy Klein processes the data on the left side of the voucher through the A/P Voucher Maintenance Process of Figure 12.1, manually posting the data on the right side of the voucher into the job cost ledger.

The A/P Voucher Update Process of Figure 12.1 increases the Outstanding Balance field of the vendor's record in the VENDORS data store by the amount listed on the voucher. It adds a record to the OPEN INVOICES data store for the amount listed on the voucher, and it posts a credit to the Accounts Payable G/L account and debits to the Expense G/L accounts listed under the heading "G/L Distribution" on the voucher. The update process also prints a journal or listing of the voucher data that has been recorded in the master data stores in case Betsy needs to review these transactions at a later date.

Horatio & Co. issues accounts payable checks twice each month. The A/P Payment Maintenance Process of Figure 12.1 begins with a review of all open invoices. The accounts payable manager selects invoices for payment and records the invoice to be paid and the amount to be paid in the PAYMENTS data store. The A/P Payment Update Process of Figure 12.1 decreases the Outstanding Balance field of the vendor's record in the VENDORS data store. It records the amount of the payment on the OPEN INVOICES data store record, and it posts a debit to the Accounts Payable G/L account and a credit to the Cash G/L account. The update process also prints a journal of the payment data, which has been updated to the master data stores.

At the end of payment processing the A/P Payment Update Process purges fully paid invoice records from the OPEN INVOICES data store and also purges the voucher records corresponding to fully paid invoices from the VOUCHERS data store. Before the voucher records are purged, the G/L distribution amounts for expense accounts are posted to the Actual Expenses field of the corresponding G/L ACCOUNTS data store records. The Actual Expenses field is used in the printing of the G/L Budget vs. Actual Reports.

Capturing Accounts Payable Job Cost Data

To understand how to change the accounts payable system to capture job cost data, it is necessary to look at the contents of the VOUCHERS data store. The VOUCHERS data store is actually made up of two files: Voucher Header and Voucher G/L Distribution. Contents forms for the files appear in Figure 12.2.

Header is a commonly used data processing term that signifies that the file contains identification information about the entities represented in the file. Header files are usually accompanied by one or more **detail** files that

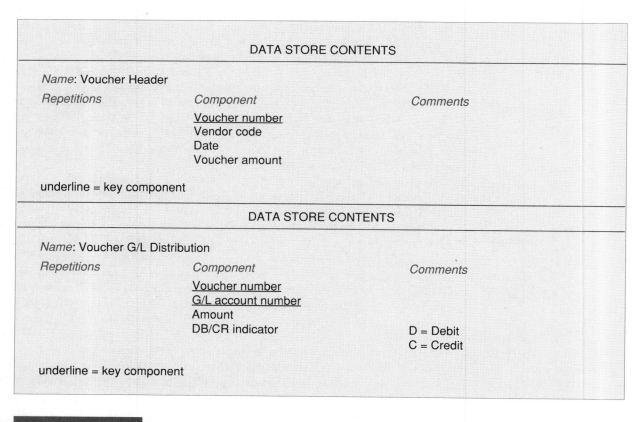

DATA STORE CONTENTS

Name: Voucher Header

Repetitions	*Component*	*Comments*
	Voucher number	
	Vendor code	
	Date	
	Voucher amount	

underline = key component

DATA STORE CONTENTS

Name: Voucher G/L Distribution

Repetitions	*Component*	*Comments*
	Voucher number	
	G/L account number	
	Amount	
	DB/CR indicator	D = Debit
		C = Credit

underline = key component

FIGURE 12.2

Voucher Header and G/L Distribution file layout.

provide further information about the entities of the header. In VOUCHERS the G/L Distribution file is a detail file that records how the full amount of the voucher is distributed to the various general ledger expense accounts.

Voucher Header and G/L Distribution make up a one-to-many relationship through the Voucher Number field. For each voucher number there is one header record in the Header file and one or more distribution records in the G/L Distribution file.

To capture the distribution of the full amount of the voucher to the various job number/account number combinations, Betsy added a second detail file called Voucher J/C Distribution to the VOUCHERS data store. J/C stands for job cost. Figure 12.3 shows a data structure diagram for the expanded VOUCHERS data store.

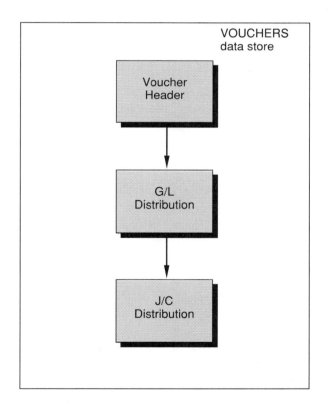

Expanded data structure diagram for VOUCHERS data store.

As a result of the addition of the Voucher J/C Distribution file to the accounts payable system, the functions of the A/P Voucher Maintenance Process had to be modified to display a J/C Distribution screen after the G/L Distribution screen is complete. Job cost data is entered through the screen and then written to the Voucher J/C Distribution file. The contents form for the new file is shown in Figure 12.4.

Validating Data Entry

The central accounting system was designed to provide validation of the vendor codes entered at the Voucher Header entry screen. If the entry is not found in VENDORS, an error message is displayed, and the user is asked to try again. If the entry is found in VENDORS, the program displays the name and address of the vendor and asks the user to confirm that the correct vendor has been selected.

The VENDORS data store is serving the role of a **reference data store** in this case. A similar validation procedure exists for the entry of G/L account numbers at the G/L distribution screen.

According to Figure 12.3, the Voucher J/C Distribution file entries should be checked against the G/L Distribution file for a matching voucher number-G/L account number combination. The addition of the Voucher J/C Distribution file to the accounts payable system also created the need to maintain additional reference files for the validation of the Job Number

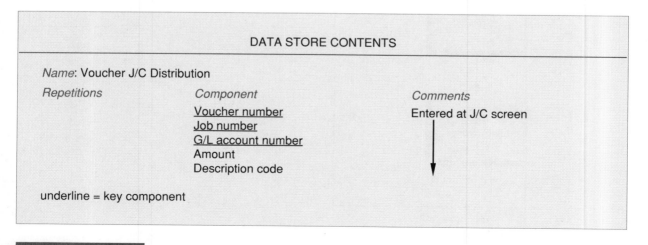

DATA STORE CONTENTS

Name: Voucher J/C Distribution

Repetitions	*Component*	*Comments*
	Voucher number	Entered at J/C screen
	Job number	
	G/L account number	
	Amount	
	Description code	

underline = key component

FIGURE 12.4

Voucher J/C Distribution file.

field and the Description Code. Of course, once a file is used, then it must be maintained, so the typical Add, Delete, Modify, Inquire, and List maintenance functions had to be developed for the Job Number field and Description Code reference files.

The Job Number reference file contained two fields: Job Number and the Active/Inactive indicator. Betsy realized she had to resolve the existence of this file with the existence of the file JOB, which was recently added to the BUDGETS data store of the cost control system.

The Description field reference file contained two fields: Description Code and Full Description. Codes are used for keyboard input to limit the number of characters the operator enters. Once a valid code is entered, the system displays the full description and asks the operator to confirm the entry.

Betsy collected several articles on the automation of job cost data that were published by the regional trade association for the construction industry. One of them included a chart of standard job cost categories with identification codes, which she used as the basis for the Description field reference file.

Batch Controls

Data entry validation through reference data stores contributes significantly to the accuracy of the data entry process. Some keyboard errors, however, cannot be detected through reference checks. For instance, the voucher dollar amounts are not checked against reference files, so an incorrect amount will not trigger a warning to the user.

Systems designers include **batch control** mechanisms in data store maintenance processes to help control the accuracy of numeric fields such as dollar amounts. Usually, the user obtains a printed report, called a **proof list** or an **edit list**, which lists and counts each transaction and which computes totals for numeric fields. The user computes the same totals manually from the original documents and compares them.

Sometimes the printed list of transactions is proofread against the source documents for added control. The term **batch control** refers to the fact that a group or a batch of transactions are being examined as opposed to the single transaction examination of reference file checks.

The A/P Voucher Maintenance Process was designed to allow the user the opportunity to review the proof list for a batch of transactions before the transactions update the master data stores. The A/P Voucher Update Process appears as an ACCOUNTS PAYABLE menu option that is activated at the user's discretion. If data entry errors are discovered on the proof list, they are corrected through the Change option of Voucher Maintenance, and the proof list is printed again to verify the effect of the change.

By contrast, the credit union system updates transactions to the master immediately after the transaction is confirmed by the teller. The daily activity log serves the purpose of the batch control proof list. If a data entry error is discovered in the log, the erroneous transaction must be reversed and the correct transaction data reentered to keep the master accurate.

As a result of the addition of the Voucher J/C Distribution file to the accounts payable system, the proof list program had to be modified to include job cost data on the printed list.

Interface with the Cost Control System

Betsy decided to let the job cost data remain in the Voucher J/C Distribution file until the corresponding invoice was paid in full and the voucher data was about to be purged from the VOUCHERS data store. At that point the modified purge program copied the job cost data to a file called the Cost Control System Interface file.

Each day, Betsy planned to transfer the Cost Control System Interface file to the microcomputer that acts as the file server for the network that runs the cost control system and to add the records of the file to the EXPENSES data store. To accomplish this, Betsy had to resolve the differences in the contents of the Cost Control System Interface file (same as Voucher J/C Distribution file, Figure 12.4—Why?), and the contents of the cost control system EXPENSES file, Figure 9.19.

The Voucher Number field of the Interface file was not needed in the cost control system, so it could be ignored. The fields for Account Number, Job Number , and Amount could be transferred intact from the Interface file to EXPENSES file. The Date and Source fields of the EXPENSES file were not contained in the Interface file, but they could be retrieved from the matching Voucher Header record and added to the Interface file record as part of the purge processing. The fields could then be transferred intact along with Account Number, Job Number, and Amount.

The Source field along with the Description field also presented a second problem. In the Interface file record the vendor and the description were represented by codes that were validated against the appropriate reference data stores at data entry time. In EXPENSES, the vendor, located in the SOURCE field, and the description, located in the DESCRIPTN field, were not represented by codes but by the corresponding full-text name or description.

This situation could not exist long before inconsistencies arose (Why?), so Betsy changed the contents of EXPENSES to include codes in the SOURCE and DESCRIPTN fields instead of full-text descriptions. This meant that the contents of EXPENSES now matched the contents of the Cost Control System Interface file except for the Voucher Number field and that a simple copying of the expanded Interface file fields except Voucher Number to the modified EXPENSES file would complete the transfer of expense transaction data from the accounting system to the cost system. Like most fourth generation environments, dBASE III PLUS provided simple field copying commands for this purpose.

The steps of the job cost data maintenance and transfer processes are summarized in Figure 12.5.

Dealing with the Reference Data Stores

The changes to the contents of EXPENSES brought the role of the reference data stores into focus. Figure 12.6 shows a new cost control system data structure diagram that emphasizes this role.

Betsy knew that Mr. Chapin planned to extend the cost control system into other departments. With this in mind she recommended that all

---

1. Betsy enters and maintains job cost data through the J/C Distribution screen now included in the A/P Voucher Maintenance Process.

2. When the invoice which a voucher represents is fully paid and about to be purged, the system automatically adds job cost data to the Cost Control System Interface file.

3. At the end of each day Betsy runs a procedure that transfers the records of the Cost Control System Interface file from the minicomputer to the microcomputer network file server.

4. Once resident on the microcomputer network file server, the transferred records are appended to the EXPENSES file, ignoring the Voucher Number field. Upon completion of the append operation, the entire cost control system database is backed up, the minicomputer Interface file is erased, and the transferred Interface file is erased from the microcomputer network file server.

FIGURE 12.5

Job cost data maintenance and transfer procedures.

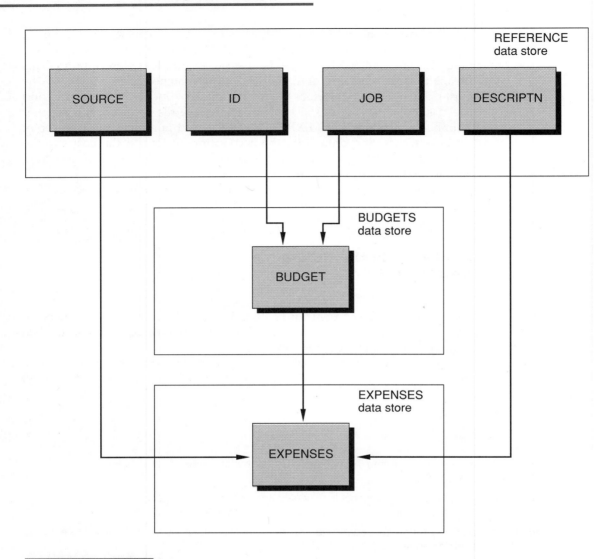

FIGURE 12.6

**A new interpretation of cost control system
reference data.**

reference data store maintenance should be assigned to a responsible
individual, carried out through the procedures of the central accounting
system, and the data transferred intact to the microcomputer-based cost
control system on a daily basis. In this way all departments access a

consistent set of reference data through file relations, and each department maintains only the budget and expense data for which it is responsible.

The accounting system files and the corresponding cost control system reference files involved in the reference data transfer procedure, along with the individual responsible for the maintenance of the accounting system files, are listed in Figure 12.7. Mr. Chapin approved the implementation of Betsy's recommendation.

Since dBASE III PLUS allows only one file relation per file at any given time, Betsy had to modify the detail line module of each inquiry and report program to change from one file relation to the next as full-text name or description fields such as Account Name, Source, and Description were output to the screen or printer.

A level 0 DFD for the expanded Accounts Payable module is shown in Figure 12.8. Similar modifications were made to the Payroll module of the accounting system. The Cost Control System Interface File regularly receives records from the Payroll module that are transferred to the microcomputer network and appended to EXPENSES along with the Accounts Payable records.

A Design Choice

The choice to copy the job cost data to the Cost Control System Interface file during the A/P Payment Update Process was made to simplify the maintenance of the Interface file and the cost control system's EXPENSES file. If a voucher data entry error is discovered after the voucher has been updated,

Responsible Individual	Accounting System	Cost Control System
Acct. Dept. Mgr.	G/L Accounts	ID
Project Mgrs.	Job Number field	JOB
A/P Supervisor	Vendors	SOURCE
Betsy Klein	Description field	DESCRIPTN

FIGURE 12.7

Reference file transfer procedure files.

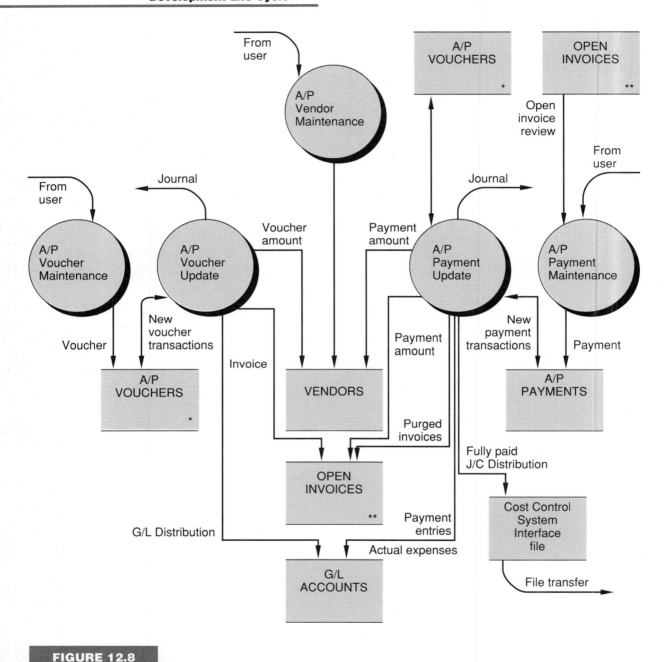

FIGURE 12.8

Level 0 DFD for expanded Accounts
Payable module.

then the corresponding VENDORS, OPEN INVOICES, and G/L ACCOUNTS records must be changed in addition to the VOUCHERS data store record. If the job cost data is copied to the Interface file after A/P Voucher Update, then the Interface file records and the EXPENSES file records would also have to be changed because of a voucher data entry error. This would involve complicated programming.

Data entry errors on paid invoices, however, are not corrected through A/P Voucher Maintenance. These errors are corrected with debit memo and credit memo transactions. Vouchers for the memos are entered through the same program as vouchers for invoices. This adds the adjusting G/L and J/C entries to the Voucher Distribution files automatically.

Debit memos are paid and credit memos are applied to invoices through the A/P Payment Maintenance Process. When the memos are purged after the update, the adjusting job cost entries are added to the Cost Control System Interface file and finally appended to the cost control system's EXPENSES file. Future reports will display both the original and adjusting entries.

This choice illustrates a common tradeoff in transaction processing systems design. When transactions that update master data stores must be changed, a design that requires a new transaction such as a debit or credit memo leads to simpler programming but more work for the user and less readable reports than a design that allows the original entry to be changed and that traces through the master data stores to make the corresponding adjustments.

Implementing the Interface

The programmer who modified the accounting system programs set up a test directory on the minicomputer for Betsy's work. Copies of the data stores represented in Figure 12.8 and the programs of the Accounts Payable module were installed in the directory. In this way Betsy's work did not disrupt the regular processing of accounts payable data.

Betsy was also supplied with simple maintenance programs for the Job Number and Description field reference files. The programs provided Add, Delete, Modify, Inquire, and List functions through a menu. Betsy would maintain the Job Number field reference file along with the Description field reference file during the prototyping phase of the cost control system development project.

For the past six months Betsy had been using a **terminal emulation** program for her accounting system data entry. A terminal emulation program allows a microcomputer to serve as a minicomputer terminal as well.

This saved Betsy the trouble of keeping both a microcomputer and a terminal on her desk.

Terminal emulation programs almost always provide **file transfer utilities**. File transfer utilities are programs that transfer data files between the microcomputer serving as a terminal and the host computer. The transfer of the Cost Control System Interface File is simple. After selecting RECEIVE FILE from the utility menu, all Betsy has to do is supply the name of the Interface file and the name of the microcomputer file that is to receive the data.

Once the additions to the EXPENSES file are resident on the microcomputer, they are added to the EXPENSES file through the dBASE III PLUS APPEND FROM command. Once the EXPENSES data store is up to date, the reference files are transferred. Since these are in ASCII format, they must be converted to dBASE III PLUS format and indexed so that the required file relations can be established.

Betsy set up a procedure to transfer the Interface file once at the end of each day. In addition to Betsy, Sam Tilden and one of the accounting system operators learned the procedure so that transfers could be accomplished in Betsy's absence.

Betsy entered several batches of test data through the new programs. She took responsibility for tracing the accuracy of the data through its entry into the cost control system, and the programmer took responsibility for tracing the data through the accounting system. Designing, coding, and testing the data entry program took five full working days for each person.

Data Conversion

With the completion of the data entry programs Betsy was ready to begin converting the manual job cost data to the automated system. In the design of the personnel component of the cost control system in Chapter 8, Betsy was assigned responsibility for transaction maintenance; Sam was assigned responsibility for master maintenance; and, most importantly, Sam and the engineers used the Reports Process in the performance of their work.

Before a transaction processing system can operate, the master data store must be in place. The BUDGETS data store of the cost control system is composed of two files: ID and BUDGET, Figures 9.24 and 9.25. Since she planned to copy the ID file from the central accounting system each day, Betsy turned her attention to the BUDGET file.

The engineering department was involved with ten jobs at the moment, so Betsy advised Sam Tilden to enter a BUDGET file record for every account number/job number combination for every month from the start of each job through the current month. Since the average age of the current jobs was twelve months, this amounted to 480 records (10 jobs × 4 G/L accounts × 12 months average = 480 records). Sam was able to complete the compilation and entry of the start-up data in the estimated two days.

In addition to the master data store, the start-up of a transaction processing system also requires all active transactions to be in place. The design of EXPENSES, the cost control system's transaction data store, was based primarily upon the manual job cost ledger, Figure 5.7. Each line of the ledger contained the job number, date, source, account number, and amount of a job cost transaction. The EXPENSES data store record, Figure 9.19, contained all these fields plus a Description field. A review of the manual ledger revealed 3700 records for the ten current engineering jobs.

Before the manual job cost ledger records could be added to the EXPENSES data store, identification codes for the Source field and the Description field needed to be retrieved and entered on the job cost ledger pages. This was due to the recent change in the EXPENSES data store to include codes for SOURCE and DESCRIPTN instead of full-text entries.

Betsy used the A/P vendor code to identify the source of accounts payable job cost ledger records, and she used the employee's social security number to identify the source of payroll job cost ledger records. The job cost category codes from the regional trade association were used to identify all descriptions.

Description data for payroll costs appeared on the weekly time logs, Figure 5.5, and the identification codes were easily transferred to the job cost ledger pages for entry into the EXPENSES data store. Sometimes a single job cost ledger entry became more than one EXPENSES record because the ledger entry represented different kinds of work performed by the same person on the same job. When this occurred, the additional records were recorded on a separate worksheet for entry into EXPENSES.

Payroll costs, however, accounted for only half of the job cost ledger entries. Description data for materials, equipment, and subcontractor costs had to be gathered from a comparison of the vendor invoices with the job cost categories. The final determination required engineering judgment. In many cases a final determination was impossible to make from the invoice data alone.

In a test run Sam was able to determine description fields for 20 A/P transactions in an hour. Since half of the 3700 existing transactions belonged to accounts payable, this meant that retrieval of the A/P description data required 90 hours of engineering time, which was not part of the original estimates made in Chapter 5. Sam assigned five engineers to the project, and it was completed in a little over two days.

The original data conversion plan called for the standardization of SOURCE and DESCRIPTN full-text entries using dBASE III PLUS. The decision to use codes for these fields changed the plan and added approximately $2500 to the cost of the project.

Entering the EXPENSES Data Store Records

The data preparation service bureau required photocopies of the job cost ledger pages, including the Source and Description field identification codes. Service bureau staff entered the data from the photocopies into ASCII files on microcomputer diskette.

The service bureau also required copies of the reference files on diskette for validation of the Account Number, Job Number, Source and Description field identification code entries. By sharing the work among personnel on three shifts, the service bureau returned a diskette containing all 3700 records to Betsy within 24 hours. The diskette file served as input to an APPEND FROM command to create the EXPENSES data store.

Processing Efficiency

Betsy was anxious to see the performance speed of the cost control system with the working database in place. She made a list of the system features whose efficiency was affected by design choices. The list included the daily file transfer procedure, transaction maintenance, printed reports, and screen inquiries. The summary table of system process considerations, Figure 9.29, is useful for recalling design choices that affect processing efficiency.

The time required to transfer files between computer systems can be estimated in advance. Betsy's microcomputer is connected to the minicomputer by a line rated at 9600 bits per second (bps). Each character in an ASCII file is represented by eight bits, and the file transfer program adds bits to the beginning and end of the character, bringing the total to ten bits transmitted for each character in the file. So a 9600 bps line should transmit 960 file characters per second (9600/10). In practice, an actual transfer rate of about 750 characters is usually achieved.

Each EXPENSES data store record is about 60 characters long, which means that approximately 12 records per second (750/60) will be transferred. Each run of the Accounts Payable Payment Update Process produces about 150 EXPENSES data store records. Each weekly run of the

Payroll module produces about 40 EXPENSES data store records. So transferring EXPENSES data store records between systems requires a few seconds of transfer time each week.

The reference files for source and description are approximately 40,000 characters. At 750 characters per second, 53 seconds (40,000/750) are required to transfer these files each day. So the amount of time spent transferring files each day is negligible, and Betsy made no effort to improve the efficiency of this operation.

Once the files are resident on the microcomputer, they must be converted to dBASE III PLUS format. New transactions must be appended to EXPENSES. Reference files must be converted and indexed. Betsy performed a test run on the full files from the dBASE III PLUS command level. All operations were completed in fifteen minutes even with other users on the network. She decided the complete file transfer procedure was adequate for her needs at the moment, but she made a note to mention that more cost control system data and/or users might degrade the performance of the file transfer procedure to the point where a new design would be required.

Transaction Maintenance

Recall Pete's and Betsy's decision to allow database reorganization upon exit from the transaction maintenance menu. Each reindex of the full transaction file took about a minute. Since Betsy was the only user of transaction maintenance, the design choice remained intact.

Reports

Betsy ran through every option of the reports menu to determine the system's response time with the working database; she was most concerned about the Budget vs. Actual reports. Each Budget vs. Actual report took a minute to prepare the required data for printing. This seemed acceptable to Betsy, but she would be sensitive to Sam's comments about the waiting time. Sam Tilden is the only user of the printed reports besides herself.

As more records are added to the database, the delay will grow. If the database levels off at 10,000 records, which was possible, the delay will be approximately three minutes for a Budget vs. Actual Report.

The reason for the delay goes back to Pete's, Sam's, and Betsy's decision not to include an update process in the design of the initial prototype of the cost control system. Consequently, Budget vs. Actual Reports must compute actual expenditures from the transactions instead of reading them from the master data store. The tradeoff involved simpler programming and file maintenance for slower report response times.

Inquiries

The benefits bought with longer file transfers and slower report response times all collected in the inquiries component, the most important and frequently used system feature. Recall Sam's second report to Mr. Chapin, Figure 8.2. The biggest and most likely system benefits came from Job Cost Reports and Monthly Budgets.

The Job Cost Reports envisioned at that stage of the project became the inquiries of the fourth prototype of the cost control system. The design choices not to summarize transactions through an update process, to replace coded fields with full-text names and descriptions upon display, and to reorganize the database often all contributed to the speed and power of the screen inquiries. Betsy was satisfied with the balance she achieved between user requirements and processing efficiency in the cost control system.

Centralized vs. Decentralized Processing

The use of computer-based information systems at Horatio & Co. is typical of many business organizations. User requirements are growing in response to modern hardware and software developments, which puts pressure on the company to expand beyond the technology of the central accounting system. Dealing with this situation in a sensible way is a serious challenge to today's managers, systems analysts, and systems designers.

The cost control system could not be developed in the third-generation environment of the central accounting system as quickly and as inexpensively as it was in the fourth-generation/CASE environment of the microcomputer network. However, the cost control system depends upon the accounting applications for most of its data, so the independent development and operation of the cost control system is not desirable. An effective means to extend the data of the central accounting system to the new environment while maintaining consistency between the environments is required.

An extended environment consisting of multiple computers and databases is usually called a **distributed processing environment**. An environment in which all users access the same computer and database is called a **centralized processing environment** [3].

One of the problems associated with a distributed environment is the incompatibility of the components. In the cost control system the modified accounting software and the daily file transfers between systems are necessary because Betsy Klein cannot access the accounting system database while running the cost control system programs. Eventually, the accounting system will complete its life cycle and be replaced. At that time Horatio & Co. will consider hardware and software environments capable of supporting both systems, but for the moment, communication between the systems at the operating system level only is a fact of life.

The extension of the accounting system data to the microcomputer network creates control problems in the accounting system. Normally, master data store maintenance programs check the effect of changes in the master on the corresponding transaction data stores. For instance, a master data store record usually cannot be deleted if matching transaction records exist somewhere in the system.

With the extension of the accounting data to another computer system, this check is no longer possible. A vendor record can be deleted in the accounting system even though the EXPENSES data store contains corresponding records because there is no way to check EXPENSES from the accounting system.

The file transfer procedure was designed to copy the full reference files each day to minimize the effects of these problems. In addition, Betsy asked Ed Henderson, the accounting department manager, not to delete obsolete vendor and G/L account records from the accounting system. Since these master data stores did not have many deletions, he agreed. In a more volatile situation this would not be possible.

The operation of a distributed environment involves managerial concerns in addition to the technical issues mentioned above. Effective communication among users avoids duplicate efforts and promotes the full utilization of existing data and software resources. This communication is the responsibility of senior management. The next chapter discusses how this communication is achieved at Horatio & Co. in relation to the newly developed cost control system.

Summary

This chapter discusses the technical issues addressed by the system builder once the user requirements have been met. To make the system prototype operational, the builder must consider the interface of the new system with existing company systems [1]. Consistency of data must be maintained. If communication between new and existing systems at the program level is possible, then the integration of the systems is accomplished through programming. If communication is possible at the operating system level only, then the integration is accomplished through file transfers and control procedures. In many cases full control is not possible.

A second consideration at this stage of the project is the conversion of manual data to the automated system. The collection and preparation of the data is time-consuming, expensive, and difficult to control. As was the case with the cost control system, data conversion effort is often underestimated in the Analysis and Design Phases of the systems development life cycle.

Once the working database is in place, the system builder considers day-to-day control mechanisms [2]. Betsy knew that full Source and Description fields could not be maintained accurately over the long run, so she changed the layout of EXPENSES to contain identification codes for these fields. She set up reference files to control the data entry of these fields, and she modified report and inquiry programs to include file relations to make full source and description values available.

Finally, the processing efficiency of the design is tested on the working database. The summary table of system process considerations, Figure 9.29, is useful for recalling design choices that affect processing efficiency.

Estimates of file transfer, data preparation, and data organization times can be calculated in advance or determined by test runs at the dBASE III PLUS command level. The ability to make realistic test runs is an important systems development capability of fourth-generation/CASE environments.

THOUGHT QUESTIONS
In Your Opinion. . .

1. What are the pros and cons of the decision not to include a file maintenance module for the OPEN INVOICES data store of the central accounts payable system? What decision would you make if you were faced with the same situation?

2. In designing the file transfer procedures, Betsy considered transferring the full-text names and descriptions directly into EXPENSES. What do you think of this idea? What are the benefits and problems associated

with this policy? Can you think of a situation in which you would choose this option?

3. Recall Figure 9.28, the data structure diagram for the initial prototype, which called for the validation of EXPENSES entries against the BUD-GET file. Do you think this can be achieved now that job cost expense transactions are entered through the minicomputer-based accounting system? Why or why not?

4. How serious does the constraint of one file relation per file in dBASE III PLUS seem to you? In light of the extra work it created, what do you think of Pete's and Betsy's decision to use dBASE III PLUS and GENI-FER to develop the cost control system?

5. What do you think about the scheduling of the daily cost control system backup after the Interface file transfer in Figure 12.5? Is there another time of day that would be more advantageous?

References

1. Jenkins, A. M. "Prototyping: A Methodology for the Design and Development of Application Systems—Part 2," *Spectrum*, Volume 2, Number 3, June 1985, pp. 1–4.

2. Kraushaar, J., and L. Shirland. "A Prototyping Method for Applications Development by End Users and Information System Specialists," *MIS Quarterly*, Volume 9, Number 3, September 1985, pp. 189–197.

3. Zmud, R. *Information Systems in Organizations*. Glenview, IL: Scott, Foresman and Company, 1983.

HORATIO&CO.

Early Implementation of the Credit Union System

The considerations of this chapter are important to Camille Abelardo, even though the software for the Horatio & Co. Credit Union System was purchased as a package. Integration with existing systems, manual data conversion, control procedures, and processing efficiency must be addressed regardless of the software development methodology.

The issue of integration with existing applications is especially complex because it extends beyond the boundary of the specific application to the organization as a whole. In exercises 1 and 2, you will see that even specialized applications like credit union account processing share data with other systems in the organization.

The difficulty of maintaining consistency among such application databases has led to the consideration of an organization-wide information architecture. The developers of the architecture consider the information requirements of the whole organization rather than the requirements of a specific application. Information architecture is discussed in Chapter 17.

Exercises

1. Does the credit union system maintain data that is also maintained by other Horatio & Co. systems? If so, identify the data.
2. Is there a need for the credit union system to communicate with other Horatio & Co. systems? If there is a need, how would the communication be implemented?
3. Develop a procedure for converting the manual credit union data discussed at the end of Chapter 3 to the automated system data discussed at the end of Chapter 2. Be sure to include accuracy controls.
4. Estimate the cost of the data conversion and the staff time required.

5. Evaluate the day-to-day control mechanisms of the credit union system.

6. Develop a set of tests to evaluate the processing efficiency of the credit union system.

CHAPTER 13

Completing the Implementation

Objectives

The development of the cost control system has reached the **operational prototype** stage [2]. The user requirements have been met by the design. The working database is in place, and processing efficiency is satisfactory. Final implementation steps involve a review of supporting procedures and personnel assignments and a look toward the future.

The specific objectives of this chapter are

1. to develop the criteria by which supporting procedures are evaluated;
2. to determine how supporting procedures are recorded and communicated to the user/management group;
3. to determine how analysis and design documents are organized into system documentation;
4. to evaluate options for the use of the operational prototype.

Finalizing System Procedures

A computer-based information system includes five components: hardware, software, data, procedures, and personnel [3]. In Chapter 8, Pete and Sam made some preliminary design choices for the procedures and personnel components. With the completion of the software and data components it is time to review those choices and complete the specification of procedures and personnel for the cost control system.

System procedures are developed to ensure the timely, accurate, complete, and secure performance of system operations. The system structure chart, Figure 9.30, can be used to perform an organized review of system operations for the purpose of finalizing procedures. The Final Analysis DFD, Figure 5.23, can be used as a road map of the context into which the automated system must fit.

Timeliness

The cost control system is simple because it contains one master data store and one transaction data store. Sam Tilden has primary responsibility for BUDGETS, the master, and Betsy Klein has primary responsibility for EXPENSES, the transactions.

Timeliness for the cost control system data is related to the monthly reports. Each month, the Budget vs. Actual Reports are used to review planned vs. actual expenditures, and the Job Cost Reports are sent to the project managers as part of the company billing process. Neither report can be delayed for any length of time.

Sam's contribution to the timeliness of the reports involves the monthly budget figures. To produce the required Budget vs. Actual Reports, a budget record for the current month must be available for every active job number/account number combination.

After reviewing the Final Analysis DFD, Figure 5.23, Sam decided to enter all the monthly budget records for a new job as part of process 6. INITIATE NEW PROJECTS. This choice avoids monthly data entry deadlines. As part of the INITIATE NEW PROJECTS process, Sam Tilden also arranged to inform the A/P supervisor and Betsy of any new vendors and job cost categories so that they could be entered into the reference data stores.

Betsy's contribution to the timeliness of the reports involves the expense transactions. Job cost transactions begin in the Accounts Payable and Payroll modules of the accounting system. Payroll checks are issued once a week, and accounts payable checks are issued twice per month. If Betsy meets these deadlines, then she automatically satisfies the requirements of the cost control system reports.

Betsy prefers to handle accounts payable and payroll transactions in batches. She accumulates payment authorization vouchers and weekly time logs until two days before checks are processed; then she enters them all at one time and files the source documents. Her alternative is to enter the vouchers and logs as they arrive, which is disruptive to her routine.

Betsy's choice to batch expense transactions affects the timeliness of the data displayed in the inquiries. If the time lag resulting from her choice is unsatisfactory to the inquiry users, Betsy will increase the frequency with which she enters transactions.

Accuracy and Completeness

The MAINTAIN BUDGETS process of the cost control system contains data validation operations and batch controls to maintain an accurate and complete master data store. Before entering the monthly budget figures for a new job, Sam combines the data from the job schedule and the purchase orders into a worksheet. The columns of the worksheet match the fields of the Budget record. After entering the monthly budget records for a new job, Sam runs the List maintenance option and proofreads his entries against his worksheet.

As a second batch control Sam runs a project-to-date Budget vs. Actual Report for the new job. The Actual column should show all zeros, and the Budget column figures should match the total budget for each account computed from the data entry worksheet. Any data entry errors, duplicates, and/or omissions will cause a difference between the printed figures and the figures computed from the worksheet.

Finally, Sam must be sure that the monthly budget figures he maintains in the cost control system match the annual budget figures maintained for his department in the central accounting system. Sam modified process 3. MAINTAIN ANNUAL BUDGETS of the Final Analysis DFD, Figure 5.23, to include a comparison of the cost control system's Year-To-Date Budget vs. Actual Report across all jobs to the Budget vs. Actual Report from the central accounting system. If the budget figures do not agree, Sam computes totals from his file of budget authorization sheets, Figure 5.10, to determine the source of the inaccuracy.

The central accounting system provides Betsy with immediate data validation as well as batch controls to ensure accurate entry of transactions. Before beginning a batch of transactions, Betsy uses accounting system inquiry programs to look up the identification codes for all source and description fields. She writes the codes on the source documents, and then she enters the transactions.

When Betsy finds a missing master data store item such as a vendor number, she asks Sam to submit a master file change form to the accounting department. The form contains the data for the missing entity. Change forms can be submitted by department managers only. They are entered by the accounting department staff, who are the only people who can change a master data store.

Once the transactions are entered, Betsy prints a proof list to match against the source documents. Data entry errors, duplicates, and/or omissions cause differences between the record count and computed totals from the proof list and from the source documents.

When all errors are corrected, Betsy updates the transactions to the masters. She stamps each source document with the date, sends the originals to the accounting department, and keeps a copy for herself. Betsy files the source documents and the final proof list in a journal of original entry for the application. During the update, a list of the transactions in the batch is printed on the printer in the accounting department. This list is filed by Accounting in case the need to trace a transaction ever arises.

Security and Control

Supporting procedures must provide protection from inadvertent and deliberate misuse of the system. Restricting access to master data stores and demanding written requests for changes signed by the responsible manager are examples of such procedures. Recording every transaction update in a printed journal also protects the security of the system.

Adequate security procedures should allow the prompt identification of irregular conditions. The system should be designed so that an independent observer or auditor can account for any condition, irregular or otherwise, through a reconstruction of the activity leading up to the condition. The auditor should be able to identify the individual responsible for each activity, and he/she should be able to reconstruct the activity without the use of documents controlled by the responsible individuals.

The managers at Horatio & Co. have a great deal of independence, but they are accountable to the project managers through the Job Cost Reports and to the general manager through the Budget vs. Actual Report. Central accounting system budget figures cannot be manipulated by the responsible manager. Changes to these figures can be done only by accounting department staff, and the changes require a budget authorization sheet, Figure 5.10, signed by both the responsible department manager and project manager.

Betsy Klein also has a great deal of independence. The company pays whatever vouchers she submits. If she does not submit it, then it does not get paid. However, Betsy is accountable to Sam through the Budget vs. Actual Report and to the accounting department, which receives a printed list from the computer system of every transaction entered on behalf of the engineering department.

Recording Procedures

Once procedures have been designed, they must be communicated to the responsible individuals. The Final Analysis DFD, Figure 5.23, provides some clues to the operation of the system, but data flow diagrams never show the sequence of operations, the timing of operations, or the decisions required for the correct performance of operations.

The best way to record procedures is to describe them by using a shorthand notation called structured English. Structured English is a combination of English and the language of structured programming. The procedure is presented as a series of sequential steps. Decisions are represented by the CASE structure, and repetitions are represented by LOOP structures.

Betsy's procedure for entering accounts payable transactions is shown as Figure 13.1.

Structured English is also used in systems analysis when several levels of the Analysis DFD leave unanswered questions about the process being modeled. Accounts payable transaction entry is performed on the morning of the 13th and the 28th of each month. Accounts payable checks are issued by the accounting department on the 15th and the 30th of each month.

Finalizing Documentation

With the completion of the procedure specifications the development of the cost control system is finished. Betsy collected the project's analysis and design documents into a file labeled "COST CONTROL SYSTEM DOCUMENTATION."

Documentation serves two purposes. Users of the cost control system can find detailed operating instructions in the documentation file, and

I.   COLLECT THE FOLDER MARKED "VOUCHERS TO BE ENTERED" FROM THE DEPARTMENT
     MANAGER'S "APPROVED TRANSACTIONS" FILE.

II. FOR EACH VOUCHER IN THE FOLDER DO THE FOLLOWING:

     A. Use the Accounts Payable Code Lookup Program to determine the identification code for the
         voucher's vendor, G/L accounts, and job cost distribution descriptions.

     B. Select the appropriate case.

         Case 1: (all identification codes found by the lookup)
             Record identification codes on the voucher.
             Hold voucher for entry into accounts payable.

         Case 2: (one or more identification codes not found)
             Prepare "Request for Master Change Form."
             Leave unsigned form and voucher in department manager's "IN" basket.

III. FOR EACH CODED VOUCHER ENTER THE VOUCHER THROUGH ACCOUNTS PAYABLE
     VOUCHER MAINTENANCE PROGRAM. COMPLETE THREE SCREENS: HEADER, G/L
     DISTRIBUTION, AND J/C DISTRIBUTION. NOTE: G/L Distribution total and J/C Distribution total
     must match the amount of the voucher entered at the header screen.

IV. DO THE FOLLOWING UNTIL AN ERROR-FREE PROOF LIST IS OBTAINED:

     A. Print the proof list from the Accounts Payable Voucher Maintenance Program.

     B. Compare the printed record count to the actual count of the vouchers.

     C. Compare the printed dollar total to the actual total of the vouchers.

     D. Proofread the G/L and J/C Distribution lines.

     E. If batch contains errors, correct erroneous voucher through the change option of the Accounts
         Payable Voucher Maintenance Program.

V.   RUN THE ACCOUNTS PAYABLE VOUCHER UPDATE PROGRAM. VERIFY BATCH RECORD
     COUNT AND DOLLAR TOTAL ON SCREEN.

VI. PHOTOCOPY VOUCHER AND FILE COPIES WITH LAST PROOF LIST IN ACCOUNTS PAYABLE
     JOURNAL OF ORIGINAL ENTRY.

VII. SEND ORIGINALS TO ACCOUNTING DEPARTMENT VIA INTEROFFICE MAIL.

FIGURE 13.1

**Written procedure for accounts payable
transaction entry.**

system builders can find a complete presentation of the analysis, design, and programming of the cost control system.

The operator instruction section contains a structured English description for every manual procedure. Betsy organized the procedures according to the processes of a general transaction processing and/or reporting system. Accompanying the procedures are structured English descriptions for the operation of the cost control system programs.

The system builders' section contains Sam's second report to the president, including a detailed budget (Figure 8.2); a complete set of Analysis DFDs and document samples (Chapter 5 figures); the Design DFD (Figure 6.11); the final Designer's Tradeoff Chart (Figure 7.4); the data store contents forms, including access paths (Figures 9.19, 9.24, 9.25); the GENIFER data dictionary entries (Chapter 10); the completed summary of system process choices (Chapter 9); the completed system structure chart (Figure 9.30 updated); the GENIFER menu, maintenance screen, and inquiry and report specifications (Chapter 10); and the dBASE III PLUS program listings (appendix).

Older versions of the systems development life cycle often included a specific documentation phase, which usually appeared at the end of the life cycle. Experience, however, has taught analysts and designers to develop documentation *throughout* the project rather than at the end of the project only. When documentation is left for the end of the project, it is usually ignored. Even if it is not ignored, it is impossible to recapture the context in which the design choices were made when one is working at the end of the project.

The Report to Management

The final report to management serves two purposes. First, it demonstrates what was accomplished in the project, and second, it sets the stage for future actions. Sam and Betsy worked together on the preparation of the report.

The report began with a recap of Sam's previous report to Mr. Chapin, Figure 8.2. The final or operational prototype met all system objectives. These objectives supported business Tactics 1–3 to varying degrees. Sam and Betsy planned to begin work on Tactic 4, reorganize the vendor base, after the presentation of this report. The budget presented in Figure 8.2 was exceeded by $2500 (10%) due to the decision to use identification codes for the SOURCE and DESCRIPTN fields, and the agreed-upon schedule of four months was satisfied.

The report continued with a section on system benefits. Sam was ready for Kim Long to transfer to another department. He volunteered to supply Mr. Chapin with three reports before the end of the year assessing the professional productivity gains and the improvements in budget control achieved by the operational prototype.

Sam related this plan to the systems development life cycle, Figure 1.3. At present, the project was in the Implementation and Evaluation Phase. Sam's progress reports would lead to a formal evaluation of the system at the end of the year in terms of the benefits outlined in Figure 8.2. This evaluation would continue throughout the following year as well.

Since Ed Henderson, the accounting department manager, worked on the project, Sam and Betsy sent him a draft of their report before submitting it to Mr. Chapin. Ed was satisfied with the report, and it was submitted to Mr. Chapin. A meeting to discuss the report was scheduled for the following Monday. In addition to Sam, Betsy, and Ed Henderson, Mr. Chapin invited Pete Willard to attend the meeting.

The Meeting with Management

Mr. Chapin was satisfied with the work of the project to date. Benefits were achieved, budgets were observed, and schedules were met. His purpose in calling the meeting was to discuss the future of the cost control system at Horatio & Co. "At this point, I think our efforts should focus on increasing the number of users of the pilot cost control system," he said. "What do you think, Sam?"

"I agree," said Sam. "At this time the Job Cost Reporting and Monthly Budgeting Modules are farthest along in the development process. These are the top-paying modules of the system requirements model (see Figure 8.2). The Engineers' Activity Analysis module consists of two inquiries that use the job cost database. The reorganization of the Vendor Base module is still just a few ideas. I am excited about the Activity Analysis module. As we expand the cost control system to other users, I want to stay involved with the development of the last two modules."

"How do we go about expanding the cost control system to other users?" asked Mr. Chapin.

Pete Willard offered his opinion. "The cost control system is one with a long life expectancy, regular utilization, and a stable environment. It is an ideal candidate for expansion to other departments. The microcomputer network can be extended in small increments to provide access to the system for a new group. The cost of adding a new group should be about the same as it was for the engineering department. Less analysis is required, but new hardware purchases are required to expand the network."

"What if the other departments cannot work with the design?" asked Mr. Chapin.

"We are approaching the last phase of the systems development life cycle, which involves the maintenance and enhancement of the cost control system," said Pete. "If necessary, we will continue to apply the prototyping cycle to enhance the operational version over time [2]."

"So Betsy will have to expand her duties as system builder and trainer," said Mr. Chapin.

"Yes," said Pete. "Someone will have to coordinate the technical side along with Ed. Betsy is the logical choice."

"Are you both willing to make that commitment?" asked Mr. Chapin.

"I'm anxious to do it," said Betsy.

"So am I," said Ed.

"What about you, Sam?" asked Mr. Chapin.

"Betsy, Pete, and I discussed this already," said Sam. "I will need some time to sit down with them to work out a realistic schedule to minimize the impact on our department."

"Fine," said Mr. Chapin. "I'll wait to hear from you before releasing a memo announcing the completion of the pilot cost control system to the entire staff. I'll invite the managers of the other departments to a demonstration of the system. Betsy, I want you to make a presentation about prototyping. The managers have been kept up to date with the progress of this project. My continued strong support and the success of the early users should encourage everyone to participate."

Summary

With the completion of the software and data components of the cost control system, Betsy turned her attention to the manual procedures that support the operation of the system. Manual procedures must meet five criteria: timeliness, accuracy, completeness, security, and control.

Scheduling inputs and processing to deliver outputs on time usually satisfies the timeliness requirement. Accuracy and completeness of data entry are supported by validation and batch control procedures. Control reports such as Budget vs. Actual Reports help to guarantee that all transactions find their way into the computer system. Finally, specific assignments of responsibilities and communication procedures between superiors and subordinates contribute to the secure operation of the system.

The activities of the Maintenance and Enhancement Phase, the final step of the systems development life cycle, vary with the system and the

operating context. With the cost control system, Horatio & Co. management
set out to extend the number of users. Enhancements to the system may or
may not occur, depending upon the needs of the new user groups. If
changes are required, they will be implemented through continued application of the prototyping cycle.

Some Final Thoughts about Prototyping

In the Job Cost Reporting and Monthly Budgeting modules of the cost
control system, prototyping was used as a software development technique. Sam Tilden, the system designer, and Betsy Klein, the system
builder, used the outputs of the early phases of the life cycle to guide the
software development process from the initial prototype through the fourth
and operational version.

At present, project plans call for the expansion of the operational prototype into other departments. Several scenarios are possible [2].

The development team may discover that an information system solution to the original problem and/or opportunity cannot deliver the anticipated benefits, and the operational prototype may be discarded. This is
unlikely in light of the experience of the engineering department, but
if software must be developed in the face of uncertain benefits, prototyping should be used because it enables the development team to make a
decision regarding the appropriateness of an information system solution
in less time and with less cost than the detailed design software development method.

The operational prototype may also evolve into the specifications for a
project to migrate computer processing to a new hardware/software
configuration that integrates accounting and job cost reporting. Such a project might be undertaken if the system is valuable but the present environment cannot provide the efficiency required for day-to-day operation, or
if the minicomputer/microcomputer configuration becomes too difficult
to manage.

A Potential Danger in the Process

With a powerful technique like prototyping, strengths can easily turn into
weaknesses if the process is mismanaged. One of the dangers of
prototyping is that a project can fall into an endless cycle of trial and

revision [1]. The techniques presented in this text can prevent this from happening.

Organizing a project into a set of measurable objectives supported by business tactics defines the scope of the project. When the objectives are met, the project is finished. Any activities that do not relate to the tactics of the project are, by definition, beyond the scope of the project. Such activities require a value analysis or feasibility review (Chapter 7) before any resources can be committed to them.

Very often, project managers will expand the scope of a project and undertake new activities because the prototyping process uncovers a need that was not anticipated. The key to managing a prototyping project is to remain flexible and open to the feedback of the process while maintaining control of the activities through frequent evaluation against the project's objectives and tactics.

Refinement Prototyping vs. Discovery Prototyping

Prototyping can be used during the early stages of the life cycle to help with the initial determination of information requirements. In fact, the development of the Engineers' Activity Analysis module employs prototyping in this way.

The systems analysis models developed for the cost control system dealt exclusively with the job cost reporting and monthly budgeting modules. No elaborate specification beyond Sam's request for some help with staff assignments was made for the Engineers' Activity Analysis module.

Based upon Sam's request, Betsy included the activity analysis BY ENGINEER, running on the EXPENSES data store, in the initial prototype. In response to Sam's feedback on the initial prototype, the analysis BY ACTIVITY was added. As Sam continues to use the analysis module, he might expand the database beyond EXPENSES. He might add new presentation and/or selection features, or he might look for a mathematical model to make staff assignments in some optimal way. At this point Sam cannot specify any details for the module because he does not know them. His specifications will grow out of his use of the evolving Activity Analysis module.

Such use of the prototyping process has been dubbed **discovery prototyping** to distinguish it from the **refinement prototyping** used for job cost reporting and monthly budgeting [1]. Discovery prototyping is a "high-risk–high-reward" prototyping option. The process is more difficult to manage, but the results in terms of effective use of the technology can be more dramatic.

In the discovery prototyping process, users educate themselves regarding the possible applications of information systems to problems and/or opportunities. The experience gained from using a working model generates specifications that would never have come to light in a "paper and pencil" interview.

On the other hand, the lack of formal specifications at the outset can cause a project to grow in scope indefinitely or to change directions many times. The management challenge to keep a project under control while maintaining an open and responsive environment is intensified in a discovery prototyping project. The project's statement of business objectives and tactics is all the more important here because it provides the only formal basis for activity evaluation.

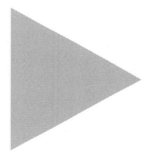

THOUGHT QUESTIONS
In Your Opinion. . .

1. Rank the four attributes of timeliness, accuracy, completeness, and security in terms of importance to the cost control system. Explain.

2. Sam Tilden decided to enter all monthly budget records for a new job at the start of the new job. What other choices can you think of for this operation? What are the tradeoffs involved in each choice? Which choice do you prefer? Why?

3. What do you think of Betsy's choices for the transaction data entry process? Why?

4. What do you think accounting department manager Ed Henderson's role is in the expansion of the cost control system to other departments?

5. What do you think of the idea of reprogramming the cost control system to run on the minicomputer? Develop advantages and disadvantages over the idea to expand the microcomputer network and evaluate the feasibility of the reprogramming project.

6. Do you think the software design methodology affects the composition of the documentation file? Do you think the software development environment affects the composition of the documentation file?

7. Review Pete's and Sam's deliberations regarding vendor reorganization in Chapter 4. Discuss preliminary choices for the hardware, software, data, procedures, and personnel components of this system.

8. If you were the president of Horatio & Co., how would you respond to the following comment from a construction project manager? "My job

is to build buildings! Let the bean counters use the computers! This cost control system is going to take me away from more important work!"

References

1. Gane, C. *Rapid System Development*. New York: Rapid System Development, Inc., 1987.

2. Jenkins, A. M. "Prototyping: A Methodology for the Design and Development of Application Systems—Part 2," *Spectrum*, Volume 2, Number 3, June 1985, pp. 1–4.

3. Kroenke, D., and K. Dolan. *Business Computer Systems*, Third Edition. Santa Cruz, CA: Mitchell Publishing Corporation, 1987.

HORATIO&CO.

Completing the Implementation

Even though the credit union software was purchased as a package, Camille needs to address all of the issues discussed in this chapter. Computer-based information systems consist of five components: hardware, software, data, procedures, and personnel. The purchase of a package eliminates the development of one component; all other requirements remain.

The four components of the systems evaluation model—timeliness, accuracy, completeness, and security—are similar to the components used to evaluate software in Chapter 11. At this point in the life cycle, the development team considers how these factors affect the final determination of supporting procedures and personnel.

The roots of the software and system evaluation models go all the way back to the Designer's Tradeoff Chart presented in Chapter 7. Timeliness, accuracy, completeness, and security were first discussed there. The specifications for these factors guided the choices made in the design and development phases of the life cycle. By collecting such items as the Designer's Tradeoff Chart and the reports to management into a file of system documentation, the development team provides a permanent means for reviewing and understanding the activities and decisions of the project.

Exercises

1. Analyze the requirements of the credit union system in terms of timeliness, accuracy, completeness, and security.

2. Rank the four attributes of timeliness, accuracy, completeness, and security in terms of importance to the credit union system. Explain.

3. Based upon your experience as a user of the Horatio & Co. credit union system, identify the specific end-of-day procedures you feel are necessary to guarantee timely, accurate, complete, and secure processing.

4. Write structured English descriptions for the procedures identified in exercise 3.

5. What analysis and design documents should be collected into the credit union system file of documentation?

6. Based upon your experience as a user of the Horatio & Co. credit union system and the procedures developed in exercise 3, make appropriate personnel assignments for system processes and procedures.

CHAPTER 14

A Critical Appraisal of the Horatio Systems

Objectives

In Chapter 7, you studied the Designer's Tradeoff Chart as a means to choose among alternative solutions to users' business problems and/or opportunities. In Chapters 8–13 you studied the Horatio & Co. systems in detail. These systems represent the respective development teams' attempt to deliver what was specified on the tradeoff chart.

Despite the concrete specifications of the tradeoff chart, systems designers still face many choices in developing the final product. This chapter reviews the Horatio & Co. systems to develop a framework for these design choices.

The specific objectives of this chapter are

1. To evaluate the design choices made in the development of the cost control system and the credit union system;

2. To compare the Horatio & Co. systems to identify similarities and differences;

3. To understand the design decision process for information systems (IS) applications and to develop a framework for design decision choices;

4. To apply the framework to systems beyond the cost control and credit union systems presented in this text.

Classifying Systems

The purpose of a computer-based system is the most important factor affecting design choices. Figure 1.2 is repeated as Figure 14.1; the categories present a useful classification of computer-based business information systems.

Both of the Horatio & Co. systems fall into the Information Systems (IS) technology category. Transaction processing and database reporting are two of the fundamental capabilities provided by IS applications. The telecommunications category also plays a role in understanding the design choices of both types of processing.

Purpose of the Credit Union System

The Horatio & Co. Credit Union System is a transaction processing system. As such, its purpose is to account for the status of its master data store entities, i.e., the members' accounts.

The system performs the clerical operations of the credit union. Its chief objectives are accuracy and efficiency. Reports and inquiries such as the Total On-deposit report, the daily activity logs, and the Account Inquiry help to ensure the accurate recording of the transaction data in the system.

Purpose of the Cost Control System

The Horatio & Co. Cost Control System has more of a managerial focus than the credit union system. The emphasis in the cost control system is on the effective use of the output rather than on the accurate maintenance of the data. One would still classify the cost control system under IS in Figure 14.1, but it would be placed to the right of the credit union system in the area of fixed-format reporting systems, decision support systems (DSS), and end-user computing (EUC).

Naturally, accurate maintenance of the data is important to Sam Tilden and the engineers who use the cost control system. For the most part, however, the maintenance functions are performed by Betsy Klein. The other

TECHNOLOGY	1960s	1970s			1980s	1990s
Information Systems (IS)	Transaction processing systems	Fixed-format reporting systems	Mini-computers	Decision support systems	End-user computing	Expert systems
Office Automation (OA)		Word-processing equipment		Electronic mail	Telecon-ferencing	Image processing
Telecommuni-cations			Decentrali-zation of EDP		Local area networking	Enterprise-wide networking
Factory Automation					Computer-integrated manufacturing	

FIGURE 14.1

Business computer systems (repeated from Figure 1.2).

users assume the data is accurate. They expect the system to organize this data into management information they can use in their engineering work.

The User Interface

The choice of user interface for an IS application depends upon the placement of the system along the spectrum from transaction processing through decision support. As a transaction processing system, the credit union system is concerned with recording transactions accurately. As a fixed-format reporting and decision support system, the cost control system is concerned with effective reporting from the database. The user interface of each system reflects these specialized purposes.

Many cost control system users opted to learn ad hoc reporting from the command level. Command-level processing represents the most flexible type of interface. It is well suited to the unpredictable and individual nature of ad hoc analysis requests.

At the command level the user can specify both the analysis to be done and the data to be included in the analysis. In return for this flexibility, the user must master the dBASE III PLUS command language. He or she is responsible for the proper selection and application of dBASE III PLUS commands for the purpose at hand. At the command level several attempts may be required before the proper report is obtained.

A menu-driven system provides a less flexible but easier-to-use interface. Menus are ideal for functions that can be identified during systems analysis, that are used repeatedly, and that are not expected to change in the future. Fixed-format reports, such as the cost control system's Budget vs. Actual Reports, and standard functions, such as the credit union system's ADD TRANSACTIONS function, are typical menu options.

If a system provides many functions, then the users may have to wade through many menus to arrive at their choice. Some users may prefer to enter the name of the appropriate program from the command level instead of choosing a menu option.

Using the command level in this way is not nearly as difficult as ad hoc reporting. Sometimes this type of processing is called "expert mode," reflecting the fact that many times users who at first are satisfied with a menu-driven system grow tired of it as they gain experience and confidence.

Command-level processing, menu-driven systems, and expert-mode systems all follow a keyboard-driven, single-screen design. The latest type of user interface, however, presents a mouse-driven, multiscreen graphic design. Multiple-screen displays presented in **windows** allow many options to be presented to the user in the form of lists. The **mouse** is used to highlight the user's choice, and selection is made by "clicking" the mouse on the highlighted choice.

The dBASE III PLUS ASSISTANT presents a simple graphic interface that does not require a mouse; the cursor control keys are used to position the highlight bar, and the Enter key is used to select a choice. The ASSISTANT is activated by typing ASSIST at the dBASE III PLUS dot prompt.

ASSISTANT users can perform command-level processing without typing commands or remembering precise syntax. The commands are organized into **pull-down menus** along the top of the screen. Use the Left-arrow and Right-arrow keys to move from one menu to the next. Use the Up-arrow and Down-arrow keys to highlight the selections within a given menu. Use the Enter key to invoke the highlighted option.

The line below the dBASE III PLUS status bar presents a list of highlight bar movement options. The bottom line of the screen displays a short description of the highlighted choice. Press the F1 key to see a more detailed description of the choice. Try it on the files of the cost control system. This type of user interface promises to be essential to the development of future computer-based systems.

At this point, you should recognize the need for different types of user interfaces, depending upon the purpose of the system. The more unpredictable and nonrepetitive the users' situations are, the more they need a flexible interface like command-level processing. The more standard and repetitive the situations are, the more likely a menu-driven interface will be suitable.

Command-level processing is more demanding than a menu-driven interface. If the purpose of the system dictates command-level processing, then the user/management group must accept the increased responsibility. If they are unwilling or unable to do so, then the system, as planned, is not feasible, and further tradeoffs are required.

Record Identifiers

The placement of an IS application along the spectrum from transaction processing to decision support also determines the way in which records are identified. Accounting system users are accustomed to identifying records through unique codes such as account number or social security number. Reporting and decision support system users are more likely to identify records through descriptive data such as a person's name.

In a transaction processing system such as the credit union system, accurate identification of the master data store record corresponding to a given transaction is critical. Otherwise, a deposit or withdrawal could be charged to the wrong account. To accomplish positive identification, transaction processing systems usually use codes such as account number to identify master data store records uniquely.

Transaction processing system data entry programs check the code typed at the keyboard against a reference file of valid codes. The reference file may be part of a master data store, or it may be a file maintained exclusively for data validation.

In the credit union system every account number typed at the keyboard is checked against the file MASTER for validity. If MASTER does not contain a record corresponding to the account number typed at the keyboard, the number is rejected and the user is asked to try again. If MASTER does contain a record corresponding to the account number typed at the keyboard, the system displays the member's name and address and asks the user to confirm that the correct member's record has been retrieved. **Context-sensitive help** is available to the user by pressing the F1 key.

In the cost control system, users of ad hoc reporting and the inquiries by SOURCE and by DESCRIPTION need to identify records in natural terms, especially when exploring the database for planning purposes.

Searching for an identifier code value such as a social security number is inconvenient and disrupts the natural question-and-answer flow of decision support system sessions.

The problem with natural record identifiers is variation. Sarah Ludwig, Sara Ludwig, SARAH LUDWIG, S. LUDWIG, and S. J. Ludwig are all possible entries for EXPENSES file records whose SOURCE is Sarah Ludwig. If all possible variations of a SOURCE field value are allowed to reach the EXPENSES file, then records that should contain the same SOURCE field value will actually contain different values, and inquiries by SOURCE will be unreliable.

Fortunately, the natural record identifier problem was solved in the cost control system. Identification codes were assigned to the values of SOURCE and DESCRIPTN to allow reference file validation at transaction entry time. At report or inquiry time, file relations between the EXPENSES file and the reference files through the code fields allowed record selection by the natural identifier.

At this point you should recognize the need for different types of record identifiers, depending upon the purpose of the system. Identification codes and reference file validation provide accurate transaction entry, while natural record identifiers provide effective reporting and decision support retrieval. Rarely will one type of identifier substitute for the other in an effective way.

The Credit Union System Data Component

As a transaction processing system, the credit union system performs many clerical operations. This purpose influences the design of the system's data component. The clerical operations of the credit union system involve the members' daily deposits and withdrawals and the monthly interest calculation.

Withdrawals present a complication because at the completion of the transaction, the members expect to leave the teller's window with the money they requested. This means that the system must have quick access to the current balance in a member's account so that members do not withdraw more money than they have in their accounts.

A member's current balance is status information about the member's account. When users need quick access to status information many times per day, systems designers usually include a status field in the master data store. The status field is updated every time a transaction occurs for a given master data store record, and when the status information is needed, the system retrieves it from this field in the master data store.

The Current Balance field of the file MASTER in the credit union system is an example of a master data store status field. Each deposit for a given account increases the Current Balance field; each withdrawal decreases the field.

The ADD TRANSACTIONS program checks this field before accepting any withdrawal transactions. The amount of the withdrawal must be less than or equal to the current balance. This process should remind you of other data validation techniques included in both systems.

The benefit of the master data store status field is quick access to the information contained in the field. The alternative is to compute the status information from the transactions every time it is needed. In transaction processing systems such as the credit union system, where the processing of the transaction waits for the retrieval of the status information, the master data store status field is a common choice.

The Cost Control System Data Component

The cost control system is not involved with operations. The Accounts Payable module of the minicomputer-based accounting system performs expense-related functions such as recording invoices and writing checks. The cost control system reports from the accounting database for control purposes and, to a lesser extent, for planning purposes.

The cost control system does not need to maintain any master data store status fields for validation purposes. The expense transactions are extracted from the central accounting system, so all expense reference data is maintained by the accounting system.

Master data store status information, however, is used in reports and inquiries as well as in data validation. The cost control system master data store entities are the account number/job number combinations. The actual monthly expenses for each account number/job number combination represent status information about these entities.

When a master data store contains only background information fields, then status information must be computed from the transactions for reports and inquiries. The decision not to use any master data store status fields, therefore, adversely affects the response time of reports and inquiries that present status information. In the cost control system these reports are the Budget vs. Actual Reports and the Job Cost Reports.

If actual monthly expenses were recorded in the Budget file for each account number/job number combination, the reports would run faster. When the prototype of the cost control system ran on the full database, user feedback concerning report response times was important. If the response times had been unacceptable, then Betsy would have had to include the status field in the Budget file and program the update, and the users would have had to take on the responsibility of maintaining a master data store status field accurately.

Since the early design of the cost control system held, the expense transactions must be retained as long as there is a chance they will contribute to the calculation of some piece of status information. Present plans call for the retention of transactions for two years after the completion of the job corresponding to the transactions. Fortunately, the cost control system transaction volume of a few hundred per month leads to between five and ten thousand transactions for a two-year period. This is not an unreasonable amount of data to store on a network file server.

Master Data Store Status
Field Tradeoffs

Unfortunately, the inclusion of a master data store status field imposes a data processing dimension upon the work of the system user. When a status field is present, someone must take responsibility for the update process. Transactions must be updated once and only once. Status information computed from the transactions must be identical to the information stored in the master data store status fields at all times. If a discrepancy arises, it must be possible to reconstruct all the necessary transactions, purged or active, to resolve the difference.

These responsibilities are difficult and time-consuming especially when one considers that a single transaction may affect more than one master. For instance, a sales transaction in an integrated accounting system usually affects the INVENTORY, GENERAL LEDGER, SALES ANALYSIS, and ACCOUNTS RECEIVABLE master data stores. Since errors are inevitable, users must expect to deal with the occurrence and correction of database errors.

At this point you should recognize that the decision to include a master data store status field in a database design depends upon the purpose of the system and the transaction volume. The credit union system required quick access to Current Balance status information, so the builders had no choice but to develop an update process, and the users had no choice but to accept the responsibility of maintaining the master data store status field accurately. If cost control system report response time remains reasonable, then Betsy and the cost control system users will be spared this responsibility.

Purging Processes

Since computers with unlimited processing and storage capacity do not exist, systems designers cannot allow the endless accumulation of transaction records and the repeated recalculation of status information. Data files must be purged eventually.

In the cost control system, purged records are copied to an archive diskette and stored. As far as the cost control system is concerned, the records never existed. This is acceptable because cost control system analyses are no longer meaningful for these records. The jobs corresponding to the purged transactions and budgets ended two years ago, so Budget vs. Actual Reports and Job Cost Reports no longer include these records. Inquiries for planning purposes no longer use these records because their age makes the cost figures obsolete. If information about a purged job is needed, the records can be restored from the archive diskette for processing.

The credit union system does not use transactions for analysis; the credit union system uses transactions to maintain status information about members' accounts. The system purges the entire transaction file once each year. For each account the credit union retains a copy of the monthly statement, which shows the opening balance for the month, all deposits and withdrawals made during the month, the interest calculation, and the closing balance for the month. Before the purge the complete database is copied to archive files and stored.

Auditability

Design choices regarding purging involve two related concepts: auditability and processing control. Auditability refers to an accurate accounting of the current state of the system. Processing control refers to the detection of processing irregularities, both deliberate and inadvertent.

Each month the credit union system accounts for its current state in the monthly statements. Each member can compare the opening balance on the statement to last month's closing balance. He or she can compare deposit and withdrawal transactions to the collected receipts, and finally, can check the arithmetic of the interest and closing balance calculations.

If a member finds a mistake in the monthly statement or if the statement does not arrive, then he or she consults with Camille Abelardo. Most likely, the member is complaining about an incorrect or missing transaction. Once the member produces the receipt for the transaction in question, Camille can check the daily activity logs for the date of the transaction to determine the cause of the discrepancy, and she can issue the necessary adjusting transactions to resolve it.

The printing of daily activity logs each day and the printing of monthly statements before the purge represent typical transaction processing system **audit trail** techniques. To audit a transaction processing system properly, one needs to be able to review all transactions in the order of original entry, and one needs to be able to determine the activity represented by the values in the master data store status fields.

For smaller systems the most common way to log transactions is with a daily printed listing or journal of original entry. Copies of the journal are usually filed in date sequence. In higher-volume systems the transactions may be logged on magnetic media such as tape or diskette to avoid generating and storing huge amounts of paper.

Since the cost control system does not maintain any master data store status fields, the audit trail requirements involve the review function only. The Expense History Reports are available for this purpose.

Processing Control

Processing control refers to a system's ability to detect and prevent such processing irregularities as keyboard errors, incorrect use of system functions, and deliberate abuse. Processing control is usually achieved through manual procedures combined with the system's audit capabilities.

A transaction processing system's maintenance programs should detect as many keyboard errors and incorrect uses as possible. For example, the credit union system transaction maintenance program does not allow a teller to enter a transaction for an account number that does not exist, and it does not allow a teller to enter a withdrawal for more than the account's current balance. The master maintenance program does not allow Camille to delete an account with a nonzero balance or an account with active transactions in the transaction file. (Try it. Remember the password is ROSEBUD.)

In the credit union system an error in the amount of the transaction cannot be detected at entry time. It will be discovered at the end of the day when a difference between the daily activity log and the cash and checks in the teller's drawer appears. Manual procedures such as reconciling the daily activity logs with the cash and checks collected during the day are called batch controls because they are designed to control the accuracy of a batch of transactions.

Some keyboard errors cannot be detected until the periodic audit. Suppose a teller applies a credit union deposit to the wrong account. No error appears at data entry time because the account number typed at the keyboard is valid. No error appears at the end of the day because the log's batch dollar total matches the cash and checks in the teller's drawer. Only when the member receives his or her statement with the deposit missing will the error come to light and be corrected.

Deliberate System Abuse

Deliberate abuse of computer systems is an important issue to all systems analysts, designers, and users. Transaction processing systems are vulnerable to unauthorized transactions and manipulation of master data store status fields.

Persons with access to a member's credit union account number can make unauthorized withdrawals that will go undetected until the monthly statement. The Horatio & Co. Credit Union issues all withdrawals over $1000 by check to make unauthorized withdrawals more difficult. Other means to protect against unauthorized transactions include membership cards and supplementary identification requirements.

Of course, the tellers have access to all account numbers, which makes both the policy of withdrawal by check and the daily activity logs seem even more important. A dishonest teller can complicate the error-detection process by discovering and using someone else's teller code for an unauthorized transaction.

An unauthorized transaction as a means to defraud a computer system suffers from the fact that the periodic audit uncovers the irregularity. A more dangerous threat comes from persons with access to a transaction processing system's master data store. With access to the credit union master, a bogus account with a current balance can be created. A normal withdrawal from the account puts real money into the hands of the thief.

Banking systems are not the only ones vulnerable to such a threat. A person with access to the data of an accounts payable system can create fictitious vendor and invoice records that might be paid if the manual control procedures do not detect them.

In the credit union system the Daily Integrity Report will detect a withdrawal from a bogus account at the end of the day. (How?) This may or may not be in time to identify the culprit. The credit union system software was designed so that only Camille Abelardo has access to the master maintenance menu. This is a sensible control for the system. Of course, if a person gets to the operating system level, anything is possible. For this reason only Camille's version of the main menu presents option 7. END, and the credit union system programs were designed so that interruption and exit to the operating system could not be achieved from inside the menu-driven system.

The credit union's small size helps auditability and processing control very much. Unauthorized transactions are less likely because Camille and the tellers know many of the members. The data files are small enough to run the Daily Integrity Report and track down discrepancies. The hardware can be physically secured and locked when it is not in use. The commercial timesharing vendor also takes steps to protect the credit union with features such as log-in procedures, data encryption, and access logs.

At this point you should recognize that auditing and processing control considerations affect the design of the software and procedures components of all systems. Systems that are close to the operations of a business and that heavily emphasize file maintenance require more control than reporting systems that assume accurate data.

Many times auditing and control choices cannot be made until the database design is complete. This illustrates the iterative nature of the systems design process. Often systems developers find themselves rethinking earlier decisions because of events that unfold later in the process. Review the systems development life cycle depicted in Figure 1.3. The feedback loops drawn in the figure represent this rethinking effort.

Updating Choices for Designs Involving Status Fields

The credit union system needs access to a member's current balance to complete a withdrawal, so the credit union software updates the master data store status fields immediately upon transaction entry. This choice guarantees that the status fields represent all transactions that have occurred.

The credit union system Transaction Maintenance module does not allow transactions to be changed. If a data entry error is made, Camille issues a reversal for the erroneous transaction and reenters the correct transaction. This choice involves more work for the user, but it provides an audit trail of every transaction, correct and incorrect, that has affected the master data store status fields. It is the price the users pay for the benefits of access to up-to-the-minute status information.

When current status information is not needed to process transactions, system designers may choose a batch update process. In such a system, transactions are written to a holding file as they occur, and the master update takes place at the end of the entry cycle.

Immediate updating depends upon indexed files for simple direct access to master data store records. The batch update design is used in environments that do not support indexed files. It is also used in environments in which not all persons who generate transactions have continuous access to the master data store. For instance, a distribution company may have sales outlets throughout the country but one warehouse from which all shipments are made. Allowing all of the outlets continuous access to the one and only inventory file would involve significant telecommunications expense that may not be justified. A less expensive solution would involve keying sales transactions into a microcomputer file and transmitting them via modem to the warehouse computer, where they are batch updated to the inventory master.

The batch update design choice allows the user to review batch controls before the transactions update the master data stores. Erroneous transactions that have not been updated can be *changed* instead of *reversed* because the error never affects the master. Some users may prefer this arrangement to the more cumbersome reversal process.

At this point you should recognize that the decision to include a master data store status field requires a choice of updating procedure. Access to current status information for transaction entry dictates immediate updating, while batch updating suffices in situations in which current access is not necessary or available. The choice of updating procedure also affects the audit trail requirements and the functions of the transaction maintenance modules.

The credit union system was the focus of this section because it contains a master data store status field. At present the cost control system does not contain a master data store status field, so these considerations do not apply.

Summary

The placement of an IS application along the spectrum from transaction processing to decision support determines many design choices, particularly those pertaining to the software and data components. The choices are summarized in Figure 14.2. If a system covers a wide range of the IS spectrum, then the design of the user interface and record identifiers may have to include both options.

In addition to user requirements, IS application designs must also provide audit trail and processing control capabilities. These considerations affect the hardware, procedures, and personnel components in addition to the software and data components.

User interface
    Command level
    Menu-driven
    Expert mode
    Graphic interface

Record identifiers
    ID codes
    Natural identifiers

Data component
    Master data store status field
    No master data store status field

Purging
    How often
    Auditability and control

Status field updating procedures
    Immediate
    Batch

Transaction maintenance functions
    Direct change
    Change by reversal

FIGURE 14.2

**Software and data component design
choices.**

Audit trail functions should provide the means to review all transactions in the order of original entry and the means to account for the values of master data store status fields. Processing control functions should provide for the prevention and correction of keyboard errors, incorrect use of system functions, and system abuse.

THOUGHT QUESTIONS
In Your Opinion. . .

1. Think about how a credit card system must be designed. Use two outputs: the monthly bill and phone authorization of purchases. From these outputs, develop contents forms for the data stores and a rough system structure chart. Make choices for the items listed in Figure 14.2

based upon your own experience with credit cards. Finally, develop audit trail and processing control functions.

2. Repeat the preceding exercise for a video cassette rental store.

3. For systems without master data store status fields, there is an alternative to removing expired records during purging. Instead of deleting the records, they are summarized into a "net" transaction. In the cost control system a purged job would be reduced to five records, one for each general ledger account. Discuss the tradeoffs involved in choosing this option.

4. Some information systems professionals see a danger in prototyping. They believe that once a system that works is developed, such considerations as processing efficiency and control, which come later in the process, will be ignored. What are your feelings? What measures can be taken to prevent this problem?

5. Think of three keyboard errors that might make their way into the cost control system Budget file and explain their effect on the Budget vs. Actual Reports and the Job Cost Reports. Is it conceivable that some errors will never be found?

HORATIO&CO.

Reviewing Credit Union System Design Choices

Chapter 7 discussed the specification of system requirements involving timeliness, completeness, security, and so on. Chapters 8 through 13 discussed system design and implementation along with the evaluation of the effectiveness of an implementation in terms of the original specifications.

This chapter identifies the specific software and data component choices that determine how well a system satisfies its Tradeoff Chart specifications. The following exercises ask you to review the choices presented in Figure 14.2 for the credit union system. Auditability, control, and transaction maintenance choices are seen to affect system security, while choices regarding status fields and updating procedures are seen to affect service to the customer and operating convenience.

Exercises

1. Act as a security consultant for the credit union. Try to invent ways to defraud the system. Use confederates inside and outside of the credit union. If you come up with a technique that evades the current auditing and control procedures, develop new control procedures.

2. Investigate the credit union's liability in cases of fraud. What about the liability of the company that developed the software? Are there ways to cover this liability with insurance?

3. How will the Daily Integrity Report detect the existence of a bogus account? How difficult does identifying the bogus account among all the legitimate accounts seem to you?

4. If the credit union system used a batch update at the end of each day instead of immediate transaction updates, what would be the effect upon service to the account holders?

5. If the credit union system used a batch update at the end of each day instead of immediate transaction updates, what would be the effect upon Camille Abelardo and the tellers?

6. Given your answers to exercises 4 and 5, what would be your update choice if you were an account holder? if you were a credit union teller? if you were the credit union system designer?

PART FIVE

Systems Development Practice—Today and Tomorrow

Part Five of this text contains three chapters. Chapter 15 discusses the range of computer-based applications beyond the Horatio & Co. systems. Chapter 16 discusses general organizational issues associated with the use and development of computer-based applications, and Chapter 17 looks ahead to the technical and organizational advances that are likely to affect systems development practice in the future.

Part Five presents the "big picture." All successful practitioners, regardless of their area of expertise, develop a broad view of their field. They use this perspective to evaluate situations and formulate plans. Without a broad point of view, systems development projects deteriorate into frustrating attempts to force previously used solutions upon new problems.

5.

CHAPTER 15

The Full Spectrum of Information Systems

Objectives

As this text draws to a close, it is time to consider the development of computer-based systems outside of the context of Horatio & Co. The cost control and credit union systems have served their purpose. They have provided specific experience with the processes of the systems development life cycle. The next two chapters put this experience into perspective.

According to Figure 1.2, both Horatio & Co. systems are IS applications. This chapter discusses IS applications in general. It identifies characteristics of the Horatio & Co. systems that are common to all IS applications, and it points out some features of IS applications that are not present in the Horatio systems. The chapter also discusses office automation, telecommunications, and factory automation technologies in light of the Horatio & Co. experience.

The specific objectives of the chapter are

1. To describe other computer-based systems and to compare them with the Horatio & Co. Cost Control and Credit Union Systems;
2. To discuss the user needs addressed by each type of system;
3. To discuss the dominant issues faced by systems development teams in the analysis, design, and implementation of each type of system;
4. To illustrate the importance of telecommunications to the design and development of computer-based systems.

Transaction Processing
Systems

In some organizations automated accounting systems have been processing transactions for more than thirty years. During this time the scope of transaction processing systems (TPS) has expanded beyond accounting to include, among other applications, sales and inventory, automated bank tellers, hotel and airline reservations, library circulation, and student registration.

Despite the increase in the complexity of transaction processing applications, the basic model presented in Figure 3.16 still applies. Every transaction processing system represents a group of entities in its master data store. Sales and inventory systems represent customers and products. Bank teller systems represent bank accounts of all kinds. Reservation systems represent rooms in a hotel or seats on an airline flight. Library circulation systems represent books and other circulating items such as records and tapes, while student registration systems represent students and course offerings.

Day-to-day transaction data is captured by a TPS and recorded in transaction data stores and/or master data store status fields. Reports and inquiries present system data to the users in a meaningful way. For example, an airline reservation system captures reservation and cancellation transactions and records this activity in the Seats Available field of the master data store representing flights. When a travel agent needs to determine whether a particular flight has any open seats, he or she uses an inquiry program to view the Seats Available field from the FLIGHTS master data store.

Implications for Systems
Analysis

Although the basic TPS model has not changed, the field has seen many new developments, particularly in the **data capture** function. Early accounting systems featured data input forms upon which the workers handling the transactions recorded essential data. The forms were collected into batches and entered into the transaction processing system by data entry personnel. Printed reports reflecting the new transactions were produced and distributed to users according to a predetermined schedule.

Modern transaction processing designs bring the data capture function directly to the site of the transaction. Supermarket and department store inventory systems, as well as library circulation systems, capture transactions at the checkout counter through the use of optical scanning equipment. The checkout clerk is entering a transaction into a TPS in the same way that Betsy entered expense transactions into the cost control system. The difference lies in the means used to enter the data and the integration of the data entry task into the actual execution of the transaction.

Data communications technology has also influenced the capturing of transactions. Betsy did not have to be in the same room as the minicomputer system to enter expense transactions. She could have been down the street or across the country. In many of today's transaction processing environments, the notions of time and distance have disappeared. If necessary, anyone can access a modern transaction processing system at any time and from any place.

For example, it is no longer necessary to go to a box office to purchase concert tickets. As long as the master data store of available seats is maintained on a computer system, anyone with a terminal, a printer, and a supply of blank tickets can execute and record sale transactions.

The individuals selling tickets at the remote sites are usually not employees of the theater; most of the time they work for a ticket service agency. Many organizations are using modern communications technology to extend the data capture function to individuals outside of the company. Modern order entry systems allow customers to enter orders directly into a supplier's computer system. Customers can check the supplier's inventory and price list and issue an order without using a single piece of paper.

Electronic funds transfer systems work in the same way. Bank accounts held by customers and companies represent the master data store entities. Customers access the system through a modem to enter their payments, or they agree ahead of time to pay a bill automatically. Payment transactions update the master by crediting the company's account and debiting the customer's account by the amount of the payment.

If the business tactics identified in the Problem and/or Opportunity Analysis Phase of the systems development life cycle call for increased efficiency in operations, then the systems analyst should consider an automated transaction processing system if one is not already in place. If an automated TPS is in place and efficiency is still unsatisfactory, then the analyst should consider the enhanced data capture and communications capabilities available through today's technology.

Implications for Systems Design and Implementation

Since automated transaction processing is over thirty years old, many standard TPS features have emerged. Whether a new TPS is being developed or an existing TPS is being replaced, the systems development team should consider the purchase of a prewritten package for the software component of a transaction processing system. This was done for the Horatio & Co. Credit Union System. It was not done for the cost control system because the reports and inquiries of the prewritten packages did not suit the needs of the engineering department personnel.

An important consideration in the use of prewritten software is the integration of the package into the existing environment. Many early transaction processing systems were designed to "stand alone," which means that each system maintained all of the data necessary for its operation. If two applications, such as inventory control and accounts receivable, used sale transactions, then each system was designed to capture and maintain sales data in separate data stores. This situation led to serious problems involving inconsistency and duplication of effort.

In addition to sharing transaction data among TPS applications, many organizations, like Horatio & Co., now use the TPS database as the basis for fixed-format reporting and decision support systems. As a result modern TPS development efforts must view transaction data at the organization level instead of the application level. Various departments may have responsibility for maintaining different portions of the data, but even in small organizations, individual departments can no longer maintain independent data stores.

The developers of the Horatio & Co. Cost Control System were careful to integrate the new system with the existing accounting applications. Although the various hardware and software components in the environment presented some technical difficulties, the small size of the organization made the work relatively simple. In large organizations the implementation of an organization-wide **information architecture** and the integration of diverse applications present one of the most challenging problems facing information systems management today [1].

Fixed-Format Reporting Systems

Usually, transaction processing systems are designed to perform tasks, and efficiency is the primary concern. A payroll application produces checks for employees and reports such as W-2 forms for the government. Automated teller machines receive deposits, dispense withdrawals, and make transfers between accounts. These systems are successful when the associated tasks are performed efficiently.

Fixed-format reporting systems do not perform tasks. Reports help managers to utilize the resources of the firm effectively [5]. For this reason fixed-format reporting systems are sometimes called **management information systems** (MIS).

The Horatio & Co. Cost Control System is basically an MIS. The Budget vs. Actual Reports help to detect cost overruns, and the inquiries provide access to historical job cost data for estimating requirements and planning work assignments.

Management information systems help managers to work effectively by transforming transaction data into meaningful information. This transformation distinguishes these systems from transaction processing systems. Very often the transformation of data into information involves **summarization** [5]. The cost control system Job Cost Reports summarize all expense transactions into one figure for each job, and all of the inquiries provide totals and subtotals in addition to the individual transactions.

MIS reports often compare summarized results to control figures to evaluate performance. The cost control system Budget vs. Actual Reports summarize expenditures into five totals by general ledger account, present the totals alongside the corresponding budget figures, and compute the difference or variance between the two. The presence of a negative number in the variance column indicates a cost overrun.

The Budget vs. Actual Report is an example of an **exception report**, the cornerstone of many MIS reporting systems [5]. An exception report provides easy identification of extraordinary situations. This is usually done by summarizing transaction data by a master data store entity identification field and comparing the summary to a predetermined control figure for each entity. Control figures are usually stored in the master data store along with background and status information.

If the master data store represents many entities, then the exception report presents only those whose performance differs significantly from the control figure. Since the cost control system Budget vs. Actual Reports represent only four general ledger accounts, comparison results for all four accounts are always presented.

Implications for Systems Analysis

It is difficult to identify a precise boundary between TPS and MIS. Systems in each category often require features normally associated with the other category. The Horatio & Co. Cost Control System is a good case in point. Its design followed the basic TPS model, Figure 3.16, but its purpose was to transform accounts payable and payroll *data* into cost control *information*.

Management information systems and transaction processing systems are not interchangeable. The detailed reports of a TPS are designed to control the accuracy of data store maintenance. The summary and exception reports of an MIS are designed to help a manager monitor the resources under his or her control. They require flexible control of data store relationships, report sequences, and data selection.

The systems analyst should use the business tactics identified in the Problem and/or Opportunity Analysis Phase of the systems development life cycle to determine the mix of efficiency and effectiveness requirements. This choice determines whether the project should provide TPS capabilities, MIS capabilities, or both.

Implications for Systems Design and Implementation

The summary and exception reports of an MIS usually draw their data from varied sources. For instance, the microcomputer-based EXPENSES data store of the cost control system is composed of transactions from the minicomputer-based Accounts Payable and Payroll accounting modules.

The cost control system was developed on a microcomputer network because of the availability of a fourth-generation/CASE development environment that was particularly well suited to the task. In many organizations current transaction processing systems were developed before the advent of fourth-generation and CASE environments, so MIS developers can expect to confront the file transfer problems of the cost control system often. The file transfer problems are compounded when the data exists on several different computer systems.

Decision Support Systems

Management information systems evolved out of transaction processing systems because TPS reports were not designed to serve managers. TPS reports support the accurate maintenance of the system's transaction and master data stores. The level of detail they provide is unsuitable for managers hoping to understand and control the processes represented by the data of the transaction processing system.

Management information systems were designed to scan the TPS database and to present concise, meaningful reports to support management activities, particularly control activities. The cost control system Budget vs. Actual Reports help Sam Tilden to control expenditures by quickly identifying jobs and general ledger accounts for which actual spending exceeds the predetermined budget.

Once Sam identifies a cost overrun, he must determine the cause. Usually, Sam uses the screen inquiries to identify the source of the overrun, and then he talks to the people involved and finally comes to a decision on whether to refuse to pay the extra amount, absorb the extra amount, or pass the extra amount along to the customer. The faster the overrun is identified, the more likely a satisfactory resolution is to occur.

Not all of the decisions Sam makes are as structured as the previous example. Consider a typical staffing decision. A new project involves 20 to 30 different tasks, which usually must be accomplished in 6 to 12 months. Naturally, Sam considers which engineers are available and whether they have the expertise required to complete the project. In the time that Sam has been department manager, available talent has never matched new project requirements perfectly.

In addition to matching the tasks to the individuals, Sam must consider qualitative factors such as professional development. If an individual possesses a unique skill, then he or she is likely to receive the same assignment over and over again. This situation is stifling for the engineer and dangerous for Horatio & Co, so every project must contain some assignments that broaden the expertise of some of the engineers. Of course, Sam must balance the need to develop his staff with the requirement to provide competent engineers who will deliver the finished products on time and within budget.

Project staffing decisions are similar to decisions made during the Evaluation of Alternatives Phase of the systems development life cycle. The decision maker must make tradeoffs among several quantitative and/or qualitative criteria. The process is usually iterative. The decision maker proposes a solution that satisfies one or more requirements, then he or she evaluates the proposed solution in light of the other requirements. Such evaluation usually produces modifications to the original idea that are then reviewed against the original requirements. The process continues until a satisfactory solution evolves.

Computer systems that support complex decision making are called **decision support systems** (DSS). DSS differ from MIS in several ways. MIS reports are used to identify problems and/or opportunities that require management attention. The problems are usually identified with an exception report that involves some quantitative measure such as dollars over budget or sales below quota. The MIS reports can sometimes provide additional information about the problem, but for the most part, the structured, predefined reports of an MIS cannot help the manager generate alternative solutions and/or arrive at a decision [2].

Decision support systems do provide a means to generate and examine alternative solutions. With the development of the Engineers' Activity Analysis module of the cost control system, Sam Tilden has begun to build a DSS for his project staffing decisions. Right now, the module provides answers to questions about past experience that Sam uses to guide his assignments. As Sam uses the system, other needs may arise. He may choose to change some of the activity analysis outputs to graphs instead of reports. He may decide to develop a mathematical analysis module that would allow him to rate each individual's suitability for each task and would then use the ratings to optimize the suitability of the assignments. He may decide to automate his assignment records to improve the determination of available resources. The choices depend upon the situation and Sam's style of approaching such situations.

Implications for Systems Analysis

When a systems analyst encounters business tactics that fall into the realm of decision support systems, the individual decision and decision maker become important. A system that supports one kind of decision very well may not work at all for another kind. A system that works well for one individual may not work for another, even if both individuals are facing the same type of decision.

The decisions attacked by DSS are usually less structured than those supported by MIS. Hence, the specific features, costs, and benefits of a DSS are harder to identify in advance [4]. Organizations should view DSS development as an investment in the effectiveness of its managers. System development and commitment of resources should be incremental. Expectations should be kept realistic, especially in situations involving qualitative criteria and data that is difficult to obtain [4].

Implications for Systems Design and Implementation

Many decision support systems involve sophisticated analysis techniques. DSS builders often use specialized software for statistical analysis, financial planning, mathematical modeling, and computer simulation. The builder must incorporate these packages into the existing hardware, software, and data environments in the same way that the cost control system builders incorporated the fourth-generation software into the existing environment at Horatio & Co.

The design of the user interface is especially important for DSS. If more than one person plans to use the DSS, then more than one interface may be required. The DSS builder should acquire only those software packages that provide several options for the user interface.

The cost control system provided command-level and menu processing. Many DSS provide processing through question and answer dialogues. In addition to reports and inquiries, most decision support systems provide graphical output. If a group of people need to use the DSS at one time, then the ability to display output on a large screen must be provided.

Expert Systems

With the development of MIS reporting and decision support, IS applications evolved from data processing into information processing systems. At the time of this writing, the first signs of the next step in the evolutionary process are emerging. **Expert systems** which process knowledge instead of information are available for several specialized applications.

In many expert systems **computer-based knowledge** takes the form of IF-THEN rules. "IF this is true, THEN that is true." A user consults with the system through a question-and-answer dialogue. During the dialogue the user supplies specific information about the problem at hand [9].

Once the user supplies information about the problem, the Expert System tries to find the answer to the user's questions in the knowledge base.

The software that allows the system to make conclusions based upon the knowledge stored in the knowledge base is called the system's **inference engine**. In addition to displaying conclusions about the user's problem, an expert system is capable of explaining how a conclusion is reached and why a particular question is being asked [2].

Implications for Systems Analysis

Systems analysts should view expert systems as another tool to bring to bear upon business problems and/or opportunities; thus the systems analyst is interested in when expert systems technology should be applied [11]. To date, expert systems have been applied to situations in which the pertinent knowledge can be translated into IF-THEN rules; the most popular field of application has been diagnosis [5].

Several medical diagnosis systems—as well as systems to diagnose problems in telephone cables, locomotive engines, and oil rigs—have made their way to market. In business, expert systems are being used to provide investment advice and to help in the preparation of income tax returns. The underlying IF-THEN nature of the knowledge in these applications is apparent.

Like decision support systems, the features, costs, and benefits of expert systems cannot be determined in advance. Initially, management should approach the technology as a research and development investment. Incremental commitments can be made as prototype versions of the system bring the feasibility of the project into focus [4].

Implications for Systems Design and Implementation

Expert system builders face a unique problem. How does one acquire knowledge from a human expert and represent it in a computer? A common method for knowledge acquisition requires the human expert to verbalize his or her thought processes while performing the decision making task. The expert system builder, sometimes called a **knowledge engineer**, tape records these sessions in an attempt to identify relevant facts and processing rules. The knowledge acquired through this process is then translated into whatever computer-based representation the knowledge engineer chooses [5].

Knowledge engineers also face many of the same problems faced by DSS builders. The hardware, software, and data environments of expert systems are not the same as the TPS and MIS environments, so effective integration of an expert system into the organization is a concern. At the time of this writing, most expert systems provide a question-and-answer user interface. As the field develops, requirements for more sophisticated interfaces will emerge.

Office Automation Systems

Many people think of office automation (OA) systems in terms of word processing software running on personal computers. To be sure, microcomputer-based word processing has had a positive impact on the productivity of office workers at all levels of the organization, but word processing is only one of many OA applications.

Office automation refers to electronic office equipment connected to a communications network [10]. Applications include **electronic mail**, which provides for the electronic storage and transmission of messages and documents, and **voice mail**, which stores and transmits spoken messages.

Teleconferencing, another OA application, consists of three applications. **Videoconferencing** utilizes full-motion or still pictures to provide sight and sound communication among geographically dispersed participants. **Audioconferencing** provides sound communication only, while **computer conferencing** enables participants to interact by using computer terminals [8].

Computer conferencing seems to have brought about a whole new means of communication. The unique feature of computer conferencing is that not all participants have to be involved at the same time. Participants can log into an on-going conference, review the previous discussion, and add their own contributions at any time.

This form of communication is known as **asynchronous communication**. Computer conferencing is the only form of teleconferencing that offers the possibility of asynchronous communication. With video- and audioconferencing, all players must participate at the same time in a synchronous manner.

Implications for Systems Analysis

Office automation did not evolve in response to a well-defined need. The technology was developed, and then an application was found [8]. Initial efforts were focused upon improving writing efficiency through word processing.

Office automation seems to be following the same path as IS applications. Attention is shifting from efficiency to effectiveness in an effort to justify the huge expenditures required to implement the technology. The problem, of course, with systems that deal with effectiveness is that specific benefits usually cannot be expressed in terms of dollars and cents [4].

Systems analysts involved in office automation should follow the same incremental commitment approaches that have evolved for the effectiveness-oriented IS applications such as decision support systems and expert systems. Efforts to identify the mix of features, responsibilities, and costs should be comprehensive. A device such as the Designer's Tradeoff Chart is especially important because of the tendency to underestimate start-up costs [10]. Since such systems involve the integration of resources across departmental lines, the participation of top management is essential.

Implications for Systems Design and Implementation

OA system builders face a unique challenge in terms of the user interface. Printed reports and screen inquiries work well for IS applications, but to a large extent, the work of the office relies upon the spoken word and face-to-face communication. At the time of this writing, many questions regarding the effective use of voice and video technology are unresolved [8].

The addition of office automation systems to a firm's collection of computer-based systems magnifies the problems of integrating the hardware, software, and data components of each. The inclusion of voice and video technology complicates matters even further. The need for such complex integration has brought about the end of the piecemeal approach to systems development [10]. Organization-wide, industry-wide, and universal standards will continue to increase their effect on the design and implementation of systems of all kinds.

Summary

While IS and OA applications were being implemented in the offices of business organizations, **computer-assisted design (CAD)** systems were being implemented in the engineering laboratories, and **computer-assisted manufacturing (CAM)** systems were being implemented in the factories. Automated design systems exist for products ranging from computer circuits to jet engines. Automated systems exist for the storage and retrieval of inventory and the movement of materials within the plant. Automatic assembly and test equipment is also available from a variety of vendors.

Attempts to integrate the business, engineering, and manufacturing systems of a firm are being made in the name of **computer-integrated manufacturing (CIM)**. Integrative efforts are underway in other industries as well. These comprehensive projects are testing the limits of our knowledge of information systems analysis, design, and management. The cultural and technical differences between production, engineering, and administrative applications are immense, but global competition continues to push the state of the art [3].

In the discussions of this chapter one underlying theme is apparent: today's systems development teams must deliver more complex systems in a shorter amount of time than their predecessors. Complex systems naturally consume more resources than simple systems, but the benefits of complex systems seem to be more difficult to identify.

As systems become more complex, systems analysts will discover the need to move higher in the organization to assemble participants for the Problem and/or Opportunity Analysis Phase of the systems development life cycle. Only organization-wide objectives and tactics can overcome the cultural and technical differences one encounters when traversing traditional organizational boundaries.

In addition to the technical problems of communication across boundaries, systems designers will face increased demands for individual customization of the user interface as the audience for today's computer-based systems continues to expand.

THOUGHT QUESTIONS
In Your Opinion. . .

1. As computer-based systems become more comprehensive and fourth-generation and CASE technologies continue to develop, what do you think is the future of prewritten application software packages?

2. As computer-based systems become more comprehensive and fourth-generation and CASE technologies continue to develop, what changes

can computer specialists expect in the roles they play in business organizations?

3. As computer-based systems become more comprehensive and fourth-generation and CASE technologies continue to develop, what do you think is the future of the systems analysis models presented in this text? Do you think data flow diagrams will be in use in the year 2000?

4. As computer-based systems become more comprehensive and fourth-generation and CASE technologies continue to develop, what do you think is the future of the systems development life cycle, Figure 1.3, presented in this text?

References

1. Brancheau, J., and J. Wetherbe. "Key Issues in Information Systems Management," *MIS Quarterly*, Volume 11, Number 1, March 1987, pp. 23–45.

2. Bullers, W., and R. Reid. "Management Systems: Four Options, One Solution," *Journal of Information Systems Management*, Volume 4, Number 2, Spring 1987, pp. 54–62.

3. Dutton, D. "In Pursuit Of CIM," *Datamation*, Volume 32, Number 3, February 1986, pp. 63–66.

4. Keen, P. "Value Analysis: Justifying Decision Support Systems," *MIS Quarterly*, Volume 5, Number 1, March 1981, pp. 1–15.

5. Kroeber, D., and H. Watson. *Computer-based Information Systems*. New York: Macmillan Publishing Company, 1987.

6. Kroenke, D., and K. Dolan. *Business Computer Systems*, Third Edition. Santa Cruz, CA: Mitchell Publishing Corporation, 1987.

7. Leigh, W., and M. Doherty. *Decision Support and Expert Systems*. Cincinnati: South-Western Publishing Co., 1986.

8. McLeod, R., and J. Jones. "A Framework for Office Automation," *MIS Quarterly*, Volume 11, Number 1, March 1987, pp. 87–105.

9. Moose, A., and D. Shafer. *VP Expert*. Berkeley, CA: Paperback Software International, 1987.

10. Sprague, R., and B. McNurlin. *Information Systems Management in Practice*. Englewood Cliffs, NJ: Prentice-Hall, 1986.

11. Weinberg, G. *Rethinking Systems Analysis and Design*. Boston: Little, Brown and Company, 1985.

HORATIO&CO.

EMPLOYEE CREDIT UNION

Wider Contexts

Only one of these exercises refers specifically to the Horatio & Co. Credit Union System; the others deal with the more general topics discussed in this chapter.

The portion of the IS application spectrum covered by the Horatio & Co. systems extends from transaction processing systems through data-oriented decision support systems. This is the area about which the most is known concerning systems analysis, design, implementation, and maintenance. Exercises 5 and 6 give you an opportunity to extend the analysis and design principles developed in this text to other decision support systems and to office automation systems as well.

The general technological framework presented in this chapter represents one context within which a systems development project must operate. Other contexts representing individual and organizational issues are presented in Chapter 16.

Exercises

1. Evaluate the changes in the procedures and personnel of the credit union resulting from the acquisition of the software package. Do you think drastic changes in procedures and personnel should be undertaken to implement a software package?

2. Think of two "stand-alone" applications that might store data about customers. Identify the problems one might encounter in an attempt to integrate the customer data from the two applications.

3. Investigate the role of the database administrator. Write a job description for the database administrator at Horatio & Co.

4. Identify a common exception report from whatever application you choose. What is the source of the transaction data? What is the source of the control item? How is the comparison made?

5. Investigate the availability of specialized software for statistical analysis, financial planning, mathematical modeling, and computer simulation at your school. Develop a DSS scenario in which one or more of these packages might be used.

6. Develop a list of main categories for an office automation system Designer's Tradeoff Chart. Compare your list with the main categories of the transaction processing and/or reporting system Designer's Tradeoff Chart presented in Chapter 7.

CHAPTER 16

Organizational Issues

Objectives

Chapter 15 described the variety of computer-based information systems in use today. The features of each type of system as well as the development issues normally associated with each type were discussed.

Systems analysts and designers use such classifications as a guide to their activities. When a project develops to the point where the features of the proposed system can be classified accurately, then past experience with systems of the same type can be used to evaluate situations and formulate plans. Such information is probably most useful in the Evaluation of Alternatives Phase of the systems development life cycle.

This chapter begins with another classification scheme. This time users are classified according to the work they normally do and the information systems they normally use. This information can guide the systems analyst in earlier phases of the life cycle such as Problem and/or Opportunity Analysis. Examining the users' work against such a framework can help to identify the proper mix of information system functions for the proposed solution.

The chapter continues with a discussion of organizational characteristics of interest to systems analysts and designers. Research has shown that an organization goes through stages in assimilating technology. Proposing information systems that are appropriate for the company's stage of growth is an important consideration in the Problem and/or Opportunity Analysis Phase of the systems development life cycle.

The chapter concludes with an analysis of the various roles played by systems analysts and designers throughout the life cycle. Depending upon the scope of the project and the organization's systems development practices, a person may be called upon to fill any combination of these roles for a given project.

A word of caution is in order here. The frameworks presented in these chapters represent the knowledge accumulated by the information systems profession over the past thirty years. They should be used as one of many tools in the analysis and design process. They are not meant to stifle creativity, intelligence, and inspiration.

New knowledge is very often gained through inspiration and a healthy disrespect for the status quo, but just as often, new knowledge evolves over time from the collective experience of the entire profession. Practicing professionals need to draw their knowledge from both sources.

The specific objectives of the chapter are

1. To classify users' work activities according to factors relevant to the development of information systems;

2. To classify the stages through which an organization goes in the assimilation of technologies such as information systems, office automation, and telecommunications;

3. To identify the roles played by systems development personnel at the various stages of the systems development life cycle.

Work Activities of Information System Users

The group of Horatio & Co. employees who participated in the development of the cost control system included Frank Chapin, Horatio & Co. president; Ed Henderson, the accounting department manager; Sam Tilden, engineering department manager; the engineers of the department; and Betsy Klein, system builder. The work activities of these individuals provide insight into the nature of administrative work in general.

Betsy Klein and the engineers perform tasks. The engineers complete projects, and Betsy builds systems. Betsy is also responsible for processing the accounts payable and payroll transactions through the central accounting system.

Consider the nature of processing Horatio & Co. accounting transactions. This is a recurring operation involving a structured set of procedures. The computer provides good support for this activity because the programs are well tested and Betsy knows how to use them. Transaction data is gathered from Sam and the engineers, and Betsy must account for the accuracy of every transaction. Betsy uses the system often. The computer-generated A/P and payroll checks are part of the operational cycle. Communication to and from Betsy is formal: written input forms, printed journals, checks, and reports.

Sam Tilden is responsible for the management of the engineering department's resources. When a new project gets underway, Sam receives an expense budget plus a schedule of required materials and services. Sam allocates personnel and equipment to the project; then he uses the Budget vs. Actual Reports to monitor progress activities.

Sam is also responsible for the professional development of the engineering staff. When making personnel assignments, he balances the requirements of the project with the need to provide interesting work for the engineers. As a project unfolds, Sam monitors each engineer's performance to make an intelligent evaluation at the semiannual performance reviews.

Consider Sam's use of the cost control system. It is regular but not as frequent as Betsy's. Budget monitoring activities are structured; project assignments are less structured. Performance evaluation is not structured at all. The computer supports budget monitoring very well; it supports project assignments less well; and it does not support performance evaluation at all. Communication to and from Sam is both written and spoken: printed Budget vs. Actual and Job Cost Reports, scheduled meetings for progress and performance reviews, and unscheduled meetings in times of crisis.

Frank Chapin, Horatio & Co. president, is responsible for the strategic management of the firm. He is concerned with the economy and its effect on the construction industry. He is concerned with competition and technological innovations in the construction trades. Mr. Chapin is also concerned with the political scene. He is active in the chamber of commerce, and serves on several fundraising committees for local and national politicians.

Frank Chapin tries to organize the information he collects about Horatio's environment into a realistic set of goals and objectives. He spends a good deal of time away from the office. He does not use a computer in any of his work. He regularly reviews several summarized financial reports from the central accounting system.

Mr. Chapin prefers verbal communication: meetings and telephone conversations. Memos to him from Horatio & Co. employees are usually no more than one page in length.

Implications for Systems Development

Several generalizations can be made from the descriptions of Betsy's, Sam's, and Frank's work. These are summarized into a collection of user profiles in Figure 16.1 [2, 10].

	Betsy	Sam	Frank
Primary Concern	Tasks	Resources	Strategy
Primary Activity	Operations	Control	Planning
Company Support	Excellent	Good	Poor
Typical Decisions	Structured	Less structured	Unstructured
Information Source	Internal	Mostly internal	Mostly external
Typical Information	Detailed	Less detailed	No detail
Time Frame	Present	Mostly present	Future
Communication	Written	Written/spoken	Spoken

FIGURE 16.1

User profiles.

This figure indicates that current computer-based information systems provide good support for operational and control activities. This is due to the facts that these activities have some structure to them and that they rely upon data generated by the firm itself. Written communication is effective because of the structured nature of the work.

The work of Frank Chapin is not well supported because of its unstructured nature. The external data that he requires is difficult to capture and standardize. While computers summarize numeric data very well, the nonnumeric data with which Mr. Chapin deals is not summarized very well by a computer.

Profiles such as Figure 16.1 are never complete. Betsy Klein's work is not entirely task oriented. She performs planning and control activities as well. Similarly, Mr. Chapin performs tasks in addition to his planning activities. The purposes of Figure 16.1 are to identify work characteristics that are relevant to the development of information systems and to identify the users most often associated with these characteristics.

The user profiles can be used to generate and evaluate ideas during the Problem and/or Opportunity Analysis Phase of the systems development life cycle. When a systems analyst encounters business tactics involving planning, then he or she should be aware that planning usually involves external data and unstructured decisions. These are difficult but not impossible conditions with which to deal. Knowing early that the proposed information systems solution is difficult rather than easy is important to the proper evaluation of the proposal.

The user profiles can also be used to detect cultural mismatches before they occur. Suppose a proposal is being made to bring the data capture function of a stock trading transaction processing system to the trading floor to increase efficiency and provide up-to-the-minute analysis of the firm's holdings. One could safely assume that many stock traders operate in Mr. Chapin's environment. Their decisions are unstructured, future-oriented, and made with highly summarized data.

The introduction of a structured, detail-oriented data capture function to such an environment could be a disaster. This does not mean the proposal should be rejected. The cultural mismatch presents a complication that can be overcome through the careful design of the procedures and personnel components of the system. The early identification of the situation allows a proactive rather than a reactive solution.

How Organizations Assimilate Technology

A useful model of the way in which business organizations assimilate information processing technology suggests a process made up of four phases [3]. Organizations progress from one phase to the next unless a disaster stalls the process.

The first phase begins with an investment in a new technology. Users are trained, and the organization undertakes one or more developmental projects. The second phase involves learning how to use the technology for tasks beyond those considered in the original plan. In this phase, users develop and refine their understanding of the technology. Phase 3 involves the development of precise controls to guide the use of the technology. The controls guarantee that later applications of the technology are more cost efficient than the first. Finally, phase 4 is achieved when the technology spreads to other groups in the organization.

During phases 1 and 3, management strives for efficiency and strong control of costs. During phases 2 and 4, management control slackens to encourage experimentation with and learning of the new technology [5]. The alternation of control periods and slack periods promotes sensible growth. A continuous period of strong control leads to stagnation, while a continuous period of slack leads to explosive growth in technology expenditures. Horatio & Co. has reached phase 4 with the minicomputer system. The system works well, and organizational learning is complete. The use of the system has spread outside the accounting department to users like Betsy Klein in the engineering department.

The database of the central accounting system is proving to be inadequate for the new planning and control applications such as the cost control system. This is typical of a phase 4 TPS. As the number of applications that share data increases, the effective management of the data resource will become a matter of great concern at Horatio & Co [5].

Horatio & Co. is at phase 2 with the microcomputer network. The original word processing applications are well established. The engineers are experimenting with computer-aided design (CAD) and mathematical analysis software; spreadsheets are everywhere, and now the cost control system is on the network.

Horatio & Co. management faces a difficult challenge. They must begin to establish phase 3 controls for the network, and at the same time promote experimentation with the cost control system to bring the cost control system out of phase 1 and into phase 2. Mr. Chapin's idea to send a clear signal to the department managers about the priority of the cost control system is a good one. In addition, he should look to slow the growth of the network until a plan and personnel are in place to handle the phase 4 network activities that are bound to occur.

Roles in the Systems Development Life Cycle

This text discusses the analysis, design, and implementation of computer-based information systems. The systems development life cycle, Figure 1.3, is presented as a blueprint for these activities. This section reviews the life cycle and presents a general discussion of the roles played by the participants during each phase.

The Problem and/or Opportunity Analysis Phase begins with the user/management group. A business problem and/or opportunity is identified, and a solution is proposed in terms of business objectives and tactics. The user/management group makes an initial judgment of the potential of an information system to support the tactics of the business plan.

The ability to make the initial judgment of the potential value of an information system is provided in several ways. In many organizations information systems professionals participate at the highest levels of management, and the use of technology of all kinds is included in the company's strategic plans. In other situations the impetus for systems development projects comes from department-level managers, such as Sam Tilden, who identify specific objectives and tactics that might be supported by an information system.

All of the systems discussed in this text have more than one user and/or use or affect data from other applications. Despite claims to the contrary, these systems cannot be developed by inexperienced persons working alone. Personal spreadsheets, financial planning models, or database query applications can be developed without professional help, but once a system leaves the realm of personal productivity, the user/management group should enlist the aid of trained information systems professionals. Such professionals do not have to be information systems specialists, and they need not work for an information systems organization. What is required, however, is the expertise and experience to deal with the technical and nontechnical issues that go beyond the use of tools and techniques.

The systems analyst(s) who join the project at this point are businesspeople who contribute knowledge of the industry and the business discipline of the application in addition to knowledge of systems development techniques [7]. At this point technical considerations serve as a loose boundary to discussions of the business objectives and tactics and the information systems functions that might support those objectives and tactics.

The Resource Allocation Role

At the conclusion of the Problem and/or Opportunity Analysis Phase of the systems development life cycle, the systems development team (user/management group + information systems professionals) evaluates the feasibility of pursuing the information systems solution further. The three feasibility criteria—operational, technical, and economic—discussed in Chapter 7 are used. A policy of incremental commitment is followed throughout the systems development life cycle. A positive decision at this point indicates a commitment to the next two phases: Build Systems Analysis Models and Evaluate Alternatives.

The decision to proceed with the project involves a commitment of resources. The authority to commit the resources may reside in a variety of places. The user/management group may control the required funds and personnel, including the information systems professionals. In such a case the development team informs whatever unit coordinates systems development for the firm of their decision to proceed.

Systems development resources in many organizations are controlled by a centralized unit. This unit is often a steering committee composed of top-level managers. The steering committee communicates with its constituency to determine an organization-wide information systems plan. New system proposals are evaluated by the committee in terms of this plan and the available resources.

Horatio & Co. displayed the makings of a steering committee. Funds were controlled by Frank Chapin, the president. Ed Henderson, the manager of the central accounting system, supplied a programmer to implement the interface between the cost control system and the accounting system. Sam Tilden supplied his time and Betsy Klein's time to develop the cost control system.

Other department managers are waiting for the results of the pilot project. If there is support for efforts beyond the implementation of the operational prototype in the engineering department, a steering committee will emerge from this group to guide such efforts.

Systems Analysis Roles

During the Model Building Phase, the Context and Analysis DFDs and the Requirements Model are developed by the team. Systems analysis tools such as data flow diagrams serve as communication and documentation devices. The DFDs represent the team's shared understanding of the current situation.

The time spent developing the DFDs depends upon the scope of the project, the number of users, the number of analysts, and the familiarity of the analysts with the users' environment. The analyst must strike a careful balance between gathering useful information about the current situation and overwhelming the team with insignificant details [8]. Team progress reviews are especially important during the Model Building Phase of the life cycle.

The process of developing the DFDs is as important to the analyst as the final product because the process provides many insights into the past experience of the user/management group and the organization as a whole. During the Model Building Phase the systems analyst tries to improve his or her understanding of the user/management group and the problem at hand by placing what he or she observes against the frameworks of system types, user profiles, and organizational growth stages. Systems analysis models provide a convenient mechanism to elicit information for this part of the analyst's work.

During the Evaluation of Alternatives Phase, the analyst becomes the facilitator of the decision-making process for the user/management group. The tradeoffs between features, responsibilities, and costs affect the hardware, software, data, procedures, and personnel components of the system. The analyst provides information, when necessary, to help the group make choices. In addition, the analyst uses his or her knowledge of the technical constraints of the situation to keep the process within realistic confines.

The Designer's Tradeoff Chart is used to help the group generate and evaluate ideas.

If the evaluation process does not produce a feasible alternative, then the development project stops. If the development team arrives at a feasible solution, then a preliminary schedule is prepared, and a proposal is made to whatever unit controls the required resources. If approved, a commitment for the first of the Design Phases of the life cycle is made.

The boundaries between the phases of the systems development life cycle are not well defined. The ending activities of one phase usually coincide with the beginning activities of the next. So it is with the Evaluation of Alternatives and the Design New System Phases. The proper evaluation of an alternative in terms of operational, technical, and economic feasibility usually requires the completion of, at least, a rough design. The same can be said for the development of the preliminary project schedule.

Design Roles

The roles of the design process and the players who fill them depend upon the software design and implementation method. The Horatio & Co. Cost Control System was developed through prototyping. Ideally, prototyping is carried out by two people [1]. One person represents the user/management group as the designer of the system features. The other person is a trained information systems professional who builds the system. The builder provides a system that satisfies the user/designer and that conforms to accepted standards of design and implementation.

If the software component is purchased as a prewritten package, representatives from the user/management group and the information systems professional staff participate in the selection of the package. A package is selected after a test run at the site where the package will be used.

The systems person is responsible for installing the package and evaluating technical features such as processing efficiency and interfaces with other company systems. The user representative is responsible for evaluating the match between the package's method of performing the work and the firm's. If the methods disagree, the user representative should consider modifying the firm's methods. If a compromise cannot be reached, then another package should be considered. The user should approach the choice of modifying a software package with caution. Software vendors produce regular updates to their products, and user modifications could eliminate compatibility with the updates.

If the software component is purchased as a package or developed through prototyping, then there is little need for a Formal Design Phase. Specifications for hardware, software, data, procedures, and personnel are fixed once a package is selected, and they evolve as the software is developed through prototyping. The Formal Design Phase remains in the life cycle to accommodate those projects that develop software through the traditional detailed design method.

The detailed design method is used when a fourth-generation/CASE environment is not available and/or feasible. The detailed design method is also used to develop software when the user/management group is not willing and/or able to participate to the extent required by the prototyping process. The detailed design process is dominated by the information systems professionals.

The detailed design process strives to produce the final version of the database and structured English specifications for every program in the system before any programming begins. Of course, full specification of all programs completely determines the procedures and personnel components as well.

Because computer programming and implementation in third-generation environments are the most expensive and time-consuming phases [9], detailed designers try to determine system requirements as accurately as possible in anticipation of the difficulty of the programming process. In addition to supporting the programming effort, the results of the detailed design provide a refined estimate of the project schedule and the required resources. A final feasibility review is usually conducted at the end of the detailed design process by the unit controlling the project resources.

Implementation and Maintenance Roles

When software is developed through the detailed design method, formal steps must be taken to bring the user back into the picture. Systems people prepare operator instruction documents and usually introduce members of the user/management group to the system through training programs.

When software is purchased or developed through prototyping, then training is less formal because the user has been continuously involved in the process. Since the user representative is the designer of a prototyped system, operator instructions usually evolve from the user's experience during the development process.

All software package vendors supply operator instructions with their products. These provide a base from which the user representative builds customized operator instructions during the evaluation and implementation work for the package.

In addition to operator instructions, technical documentation for the system builder is usually collected at the end of the implementation phase. The techniques presented in this text are designed to provide technical documentation throughout the process, no matter what method is used to acquire the software. Fourth-generation programs and specifications for report writers, screen painters, and data dictionaries automatically provide much of the technical documentation that previously had to be maintained on paper.

Technical and operational maintenance roles are considered during the design of the system and finalized at the end of system implementation. Company management should be concerned with becoming dependent upon one person for the successful maintenance of the system. Maintenance plans should always include redundant assignments. The high turnover among computer-literate workers of all kinds makes this goal difficult to achieve.

Information Centers

Betsy Klein contributed to the development of the cost control system long before she built the software for it. Her expertise promoted interest in microcomputing among members of the engineering department. That interest eventually provided the impetus for the cost control system development project. As the use of the cost control system spreads, it is easy to imagine Betsy taking on responsibility for introducing information systems technology to other departments. These activities could eventually occupy all of her worktime.

Many companies are organizing groups of workers like Betsy into units known as **information centers**. Information centers support the development of applications by the user/management group. These centers conduct training in the use of computing productivity tools, implement data access mechanisms, consult with users to help solve their problems, and establish communication among users and between users and the central information systems organization [6].

The boundary between the central information systems organization and the information center varies with each company. In some organizations information center personnel are providing the kind of analysis and design services that Pete Willard provided to Horatio & Co. In other organizations a project is transferred to the central information systems organization when its design begins to affect the corporate database. The independence with which an information center operates is a good indicator of the quality of the data resource management in the firm.

Current Practices

A recent survey on current practices in systems analysis and design [4] reports the following:

Of the organizations responding to the survey, 71% are using or are considering using a systems development life cycle of some kind.

70% are using or are considering using structured analysis and design methods.

62% are using or are considering using prototyping.

80% are using or are considering using CASE tools such as program generators, report writers, screen painters, and automated data dictionaries.

87% are using or are considering using information centers.

The authors of the study make several predictions regarding future trends [4]:

Organizations will continue to develop their computer-based information systems within the framework of a systems development life cycle.

Structured approaches will increasingly supplement traditional/classical approaches to develop computer-based information systems.

Automated aids will be used increasingly to support specific systems analysis and design tasks.

Prototyping will be used increasingly to facilitate the definition of the users' requirements in a computer-based information system project.

Information centers will be increasingly used to support the development of computer-based information systems by end-users.

THOUGHT QUESTIONS
In Your Opinion. . .

1. Think about two decision support systems: one to help select next semester's courses and one to help select your first job after graduation. Evaluate each system according to the categories presented in Figure 16.1.

2. What do you think would be the most difficult part of developing the course selection system? the job selection system?

3. Specify your ideal user interface for the course selection system. Do the same for the job selection system.

4. Evaluate the assimilation of academic computing technology at your institution. Identify the various technologies in use. For each technology, identify the phase that the institution has reached. Are the controls proper for the phase? Are growth and experimentation proper for the phase? If not, what suggestions would you make?

5. What do you think will be the impact of CASE technology on the systems development roles presented in this chapter? Which roles will be affected most? Why? Which roles will be affected least? Why?

6. What is your impression of the survey of current practices presented at the end of this chapter? Were you surprised by any of the results? Do you agree or disagree with the predictions regarding future trends? Why?

References

1. Jenkins, A. M. "Prototyping: A Methodology for the Design and Development of Application Systems," *Spectrum*, Volume 2, Number 2, April 1985, pp. 1–8.

2. Kroeber, D., and H. Watson. *Computer-based Information Systems*. New York: Macmillan Publishing Company, 1987.

3. McKenney, J., and F. W. McFarlan. "The Information Archipelago-Maps and Bridges," *Harvard Business Review*, Volume 60, Number 5, September-October 1982, pp. 109–119.

4. Necco, C., C. Gordon, and N. Tsai. "Systems Analysis and Design: Current Practices," *MIS Quarterly*, Volume 11, Number 4, pp. 461–476.

5. Nolan, R. "Managing the Crisis in Data Processing," *Harvard Business Review*, Volume 57, Number 2, March-April 1979, pp. 115–126.

6. Sprague, R., and B. McNurlin. *Information Systems Management in Practice*. Englewood Cliffs, NJ: Prentice-Hall, 1986.

7. Vitalari, N. "Knowledge as a Basis for Expertise in Systems Analysis: An Empirical Study," *MIS Quarterly*, Volume 9, Number 3, September 1985, pp. 221–241.

8. Yourdon, E. "What Ever Happened to Structured Analysis?" *Datamation*, Volume 32, Number 6, June 1986, pp. 133–138.

9. Whitten, J., L. Bentley, and T. Ho. *Systems Analysis and Design Methods*. St. Louis, MO: Times Mirror/Mosby College Publishing, 1986.

10. Zmud, R. *Information Systems In Organizations*. Glenview IL: Scott, Foresman and Company, 1983.

HORATIO&CO.

Organizational Issues

As in Chapter 15, the first exercise refers to the Horatio & Co. Credit Union System, and the remaining exercises refer to the more general concepts presented in this chapter. All of the chapter concepts—users' work activities, technology assimilation, and systems development roles—are covered in the exercises.

In exercises 2 through 5, you will be asked to make observations about applications and organizations. If you have full-time or part-time work experience, use it for these exercises. If you do not have work experience, use your institution and its systems—registration, library, etc.—or your own experience with retail sales, credit card, and banking systems as the source of your observations.

Exercise 6 again points out the importance of a competent systems builder like Betsy Klein to the success of prototyping projects such as the cost control system.

Exercises

1. Analyze Camille Abelardo's and Walter O'Reilly's work for the credit union according to the framework presented in Figure 16.1.

2. Describe a cultural mismatch, real or imagined, similar to the stock trader example presented in the text. Propose a solution to the mismatch.

3. Describe a technology implementation, real or imagined, that suffers from stagnation due to a continuous period of strict control.

4. Describe a technology implementation, real or imagined, that suffers from explosive growth in technology expenditures due to a continuous period of slack controls.

5. As an organization assimilates a technology, is the formation of a steering committee inevitable? Once formed, is it permanent?

6. Suppose Betsy Klein did not work for Horatio & Co. Develop two different scenarios for the assimilation of IS technology in the organization.

CHAPTER 17

A Look at What's Ahead

Objectives

This chapter considers two trends that are likely to have significant impact on systems development practice in the near future. The first is the continued development of computer-aided software engineering (CASE) technology, and the second is the concept of information as an organization-wide resource.

The specific objectives of this chapter are

1. To examine the full range of CASE capabilities and the emerging notion of an organization-wide information architecture;
2. To determine the effects of the above on the work of the systems development process.

CASE Technology

Throughout this text you have seen how various groups within an organization use information systems technology to advantage. Financial managers use automated accounting, control, and analysis systems to guarantee efficient operations and effective management. Administrative workers of all types use office automation and telecommunications to dispel the notions of time and distance, while manufacturing groups use factory automation technology to design and build better and less costly products.

In the past, information systems organizations have not kept pace with their counterparts in the creative application of information systems technology. The analogy of the shoemaker's children who go barefoot because their parent is busy building shoes for everyone else is often used to describe this situation.

The situation, however, is changing rapidly. CASE products are appearing at an ever-increasing rate. The products range from specialized task-specific tools to integrated systems that support the entire systems development life cycle.

The following sections describe the capabilities available in the CASE marketplace today and the effects these capabilities are likely to have on the systems development process. The list of products presented here is representative and by no means complete. See [7] for a comprehensive analysis of the CASE marketplace.

Documentation Maintenance

Automated documentation maintenance is the distinguishing characteristic of CASE products [7]. GENIFER's data dictionary allowed Betsy to maintain facts about the cost control system's data stores [4]. GENIFER's Menu, Screen, and Report Maintenance modules recorded facts about other system objects such as data flows and system structure chart modules. The maintenance functions are the same as those of any other transaction processing system: Add, Change, Delete, Inquire, and List.

System developers, however, use graphical models in addition to the text-oriented data dictionary. You already know that data flow diagrams and system structure charts are tedious to draw and maintain manually. Today software tools are available for this purpose. Using a mouse or similar device, the user selects a symbol from a list of choices, places the symbol at the desired location, and indicates which of the other symbols on the diagram should be connected to the new symbol. Text for identifying the symbols, of course, is still entered in the traditional way.

Figures 17.1 and 17.2 show the cost control system Context and Analysis DFDs drawn with one of these tools. Compare these with the hand-drawn figures you did in Chapter 5. Which are easier to read? Which would you rather develop and maintain? Do you think computer-generated figures will replace hand-drawn figures completely?

The efficiency of the diagramming software allows the analyst to keep graphical documentation current. It is difficult to muster the motivation to spend many hours redoing hand-drawn documents to keep pace with a development project. In an automated environment, changes to DFDs and other diagrams are made as easily as changes to a word processing document.

Although diagramming efficiency is important, the most significant advantages of automated documentation are achieved by the tools that integrate text-oriented dictionaries and graphics-oriented diagrams and charts. The most popular of these packages is EXCELERATOR from Index Technology [2].

The integration of text and graphical entries provides consistency. In these environments it is impossible to identify a data store with one name on a DFD and another name in the database. It is impossible to lose track of a DFD symbol or create dictionary entries for undefined objects. The software maintains consistency among documentation objects at all times.

In addition to automatic consistency checks, integrated documentation environments provide flexible access to the system design database. This encourages frequent and effective review of the documentation by the user. Easily available cross-reference reports and object inventory reports help to control errors and omissions.

The concepts of the preceding paragraphs should be familiar. The same techniques are used to maintain the accuracy, consistency, and completeness of the cost control and credit union databases. In essence, these CASE tools use the organizing powers of the computer to build and maintain a system design database in the same way that IS applications use the organizing powers of the computer to build and maintain transaction processing databases. The entities represented by the system are different, but the file maintenance and control concepts are exactly the same.

Programmer Productivity Tools

GENIFER's code generator is a programmer productivity tool [4]. Instead of developing the logic and writing the steps of menu processing, file maintenance, and reporting, GENIFER users describe the fundamental features

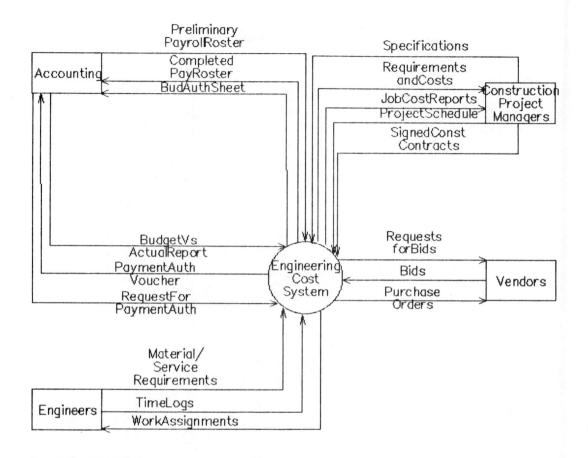

FIGURE 17.1

The cost control system Context DFD.

of the process, and GENIFER generates the program from these specifications. A programmer can describe many processes by using a code generator in the same time it takes to design, code, and test a single program.

Programmer productivity tools work with common data processing tasks. The designers of a productivity tool identify the structure of the task and supply an interface through which the programmer describes his or her specifications for the structure. Report writing, screen painting, data description, and discovery prototyping are the most popular tasks supported by programmer productivity tools [1, 2, 3, 4, 5].

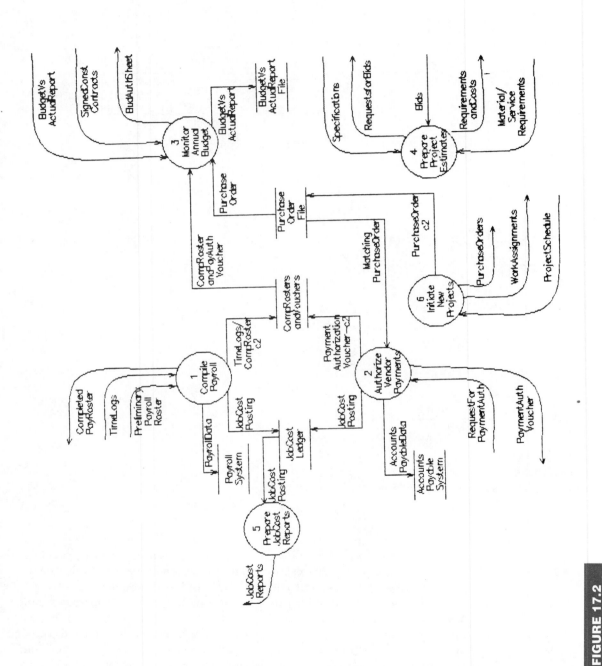

FIGURE 17.2

The cost control system Final Analysis DFD.

Project Management Tools

Many integrated CASE environments provide project management modules. These are similar to stand-alone management products that are used for all types of task-oriented projects. The advantage of the integration of the project management module with other CASE modules comes from the data that can be shared among the components. The use of shared data reduces duplicate effort, redundancy, and inconsistency in maintaining the systems development database.

FOUNDATION from Arthur Andersen & Co. [3] advertises a project-specific database. The entities represented by the system are the project tasks. Reference information regarding skill and time requirements for the tasks is maintained. A report combining tasks and requirements is used in conjunction with a list of available project team members to determine a time schedule and staff assignments. As the project unfolds, reports comparing estimated and actual times to complete the project tasks are available, and revised estimates can be used to update the project schedule. Once the system is implemented, the project management module can be used to maintain accurate records of changes made to the system.

The use of project management software varies from one individual to the next. The effort required to capture, record, and maintain the required data is great, even for a small project. Some project managers will choose to maintain all of the necessary data in return for the accuracy and control of a computer-based system. Other managers will maintain smaller databases and use the system to manage projects at a less specific level.

Problem Analysis Tools

The Problem and/or Opportunity Analysis Phase of the systems development life cycle has received the least attention from CASE system developers. This is not surprising. Problem analysis is unstructured, and to a large extent it depends upon external data that is difficult to capture and represent in a computer system.

The Planning Workstation from KnowledgeWare Inc. [5] advertises the ability to gather data about the goals of the enterprise, the organizational structure, the geographic structure, how business functions are currently performed, and candidate development projects. The data is analyzed through a series of diagrams, tables, and matrices.

Automated Design Tools

Advances in artificial intelligence are being applied to the systems development process. Many CASE systems are capable of enforcing DFD rules such as the conservation of data flows, while others can help normalize a database from the data dictionary entries [7].

CASE tools are also beginning to address the more creative design tasks. The Bachman Product Set from Bachman Information Systems [1] advertises expert advisory systems for systems analysis, database design, and program generation. The designer can use the systems to review his or her design for flaws, or the designer can allow the systems to generate the first version of the design, which can then be modified, if necessary, according to the particular situation.

Reverse Engineering

The benefits of CASE are of little use to an information systems organization that spends most of its time and money maintaining existing systems. It is not realistic to expect these organizations to prepare logical and physical models of an existing system for entry into a CASE design database. In many cases the task may be impossible because the original system developers have moved to other positions, and no one in the organization completely understands the underlying logic of the system.

The Bachman Product Set [1] address this difficulty through a concept known as **reverse engineering**. Reverse engineering modules analyze the programs of an existing system to generate the system design database automatically. The database includes logical and physical models that can be presented in both graphical and text format. The specifications exist just as if they had been entered by an analyst in a new systems development effort.

Effects of CASE on Systems Development Practice

CASE technology is sure to follow the same path as other information processing technologies. Early applications concentrated on efficiency through task-specific tools. As these tools are integrated into comprehensive systems, the effectiveness of the organization is improved.

An integrated set of CASE tools provides development team members, present and future, with a common point of focus: the system design database. The productivity of the individual in the development portion of the life cycle is improved, and reliance upon individuals in the maintenance portion is decreased. Achievement of these goals will eliminate some of the major difficulties of information systems development as it is practiced today.

Information Architecture

As an organization accumulates an applications portfolio, the need to integrate new applications with a growing base of existing systems places additional demands upon the systems development process. One way these demands are being addressed is through the concept of an information architecture [6].

An information architecture is a high-level model of the information requirements of the organization. The model affects the systems development process by accelerating and constraining the consideration of the data component of the new system. In the future, system developers may very well find themselves looking to the system's database and its interaction with the organization-wide architecture to determine the design of system functions and processes.

Effects of Information Architecture on Systems Development Practice

The response of the systems development profession to the demands of an organization-wide information architecture has been to develop **data-oriented modeling** techniques to supplement **function-oriented** techniques such as data flow diagramming.

A model of the organization-wide information architecture can be built by combining data-oriented models for existing systems into a global model. Comparison of a new system data model with the global model reduces inconsistency and redundancy and provides insight into system functions regarding data collection, maintenance responsibility, and security.

In organizations that have implemented an information architecture, systems analysts routinely develop logical models of the current and new systems' data in addition to the Analysis and Design DFDs. Development of the physical model of the new system proceeds from a consideration of both the data and function models [9]. Emphasis upon one or the other varies according to the situation and the organization.

The Entity-Relationship Model

The **entity-relationship (E-R) model** is a commonly used data-oriented modeling technique [8]. To develop the E-R model for a given application, the analyst uses the system objectives developed in the Problem and/or Opportunity Analysis Phase, Figure 4.2, to build a mental picture of what needs to be included in the system database.

From the mental picture the analyst generates a list of **entity types**, i.e., a list of "things" of interest to the business and/or the system that need to have data stored about them. A list of entity types for the cost control system is shown in Figure 17.3.

Upon completion of the list, the analyst considers all pairs of entity types and the **relationship type**, if any, that exists between them. The relationship type is described by the role that each entity type plays in the relationship and the number of entity types that can participate. For example, each Engineer performed one or more Activities, and each Activity was performed by one or more Engineers. Figure 17.4 shows a list of all relationship types for the cost control system entity types.

One of the advantages of the E-R model is its ability to provide information about the meaning of the data [8]. For example, the relationship between Engineer and Activity connotes Sam's need to see a listing of what

Expense
Budget
Job
G/L Account
Vendor
Engineer
Activity

FIGURE 17.3

Entity types for the cost control system.

Each Expense is charged to one, and only one, Budget.
Each Budget is charged with one or more Expenses.

Each Expense is charged to one, and only one, Job.
Each Job is charged with one or more Expenses.

Each Expense is charged to one, and only one, G/L Account.
Each G/L Account is charged with one or more Expenses.

Each Expense comes from one, and only one, Vendor or Engineer.
Each Vendor or Engineer generates one or more Expenses.

Each Expense represents one, and only one Activity.
Each Activity generates one or more Expenses.

Each Budget is tied to one, and only one, Job.
Each Job has one or more Budgets.

Each Budget is tied to one, and only one, G/L Account.
Each G/L Account has one or more Budgets.

Each Engineer performed one or more Activities.
Each Activity was performed by one or more Engineers.

FIGURE 17.4

Cost control system relationship types.

activities were performed by a given engineer. If Sam were developing a system to make personnel assignments, he might say, "Each Engineer *can* perform one or more Activities." Thus the E-R model communicates the sem**antics** of the situation better than a data flow diagram containing two data stores marked ENGINEER and ACTIVITY.

The final step in the development of the E-R model involves combining the text specifications of entity types and relationship types into a graphical model, Figure 17.5. Compare this model to the data structure diagram for the cost control system database shown in Figure 12.6. They are remarkably similar.

Data-Oriented vs. Function-Oriented Modeling

Notice that the E-R model identified the need for the JOB reference data store immediately, while Sam and Betsy did not realize this need until after

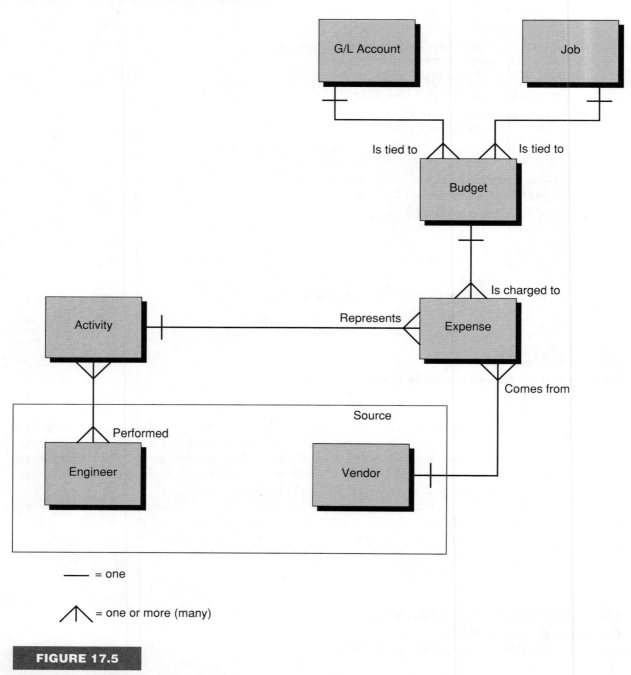

FIGURE 17.5

Entity-relationship model for cost control system.

the test of the initial prototype. This is to be expected since E-R is a data-oriented modeling technique.

The strength of the function-oriented data flow models lies in their ability to define the scope of the new system clearly and to identify interfaces within the new system and between the new system and the existing infrastructure. Current practice calls for the use of both techniques.

Requirements of a Successful Systems Analyst

Turn to page 4 of this text and review the list of qualifications for a career in systems development. The need for every one of these qualifications has been demonstrated in this text, but the emphasis has been on the technical and economic items such as software engineering, databases, and efficiency.

This emphasis is appropriate for a first undergraduate course in systems analysis and design. In fact, other technical courses such as database management will enhance your qualifications as a systems analyst.

You can improve your qualifications in the less technical areas by taking elective courses in interpersonal relations, organizational development, psychology, sociology, political science, writing, speaking, acting, or any of the functional areas of business.

No matter how extensive your classroom preparation, however, it will not complete your education. Systems development is a craft, and as such, it requires practice. Experience is a wonderful teacher.

Which brings us to possibly the most important qualification for a career in systems development: a lifelong commitment to learning. This text began with a quote from Gerry Weinberg about the education of a systems analyst. After coming this far, your belief in that statement should be stronger than it was when you first read the words on page 4.

Good luck!

References

1. "THE BACHMAN APPROACH." Cambridge, MA: Bachman Information Systems, 1988 (617-354-1414).

2. "A GUIDED TOUR OF EXCELERATOR." Cambridge, MA: Index Technology Corporation, 1987 (617-494-8200).

3. "FOUNDATION." Chicago: Arthur Andersen & Co., 1987 (312-580-5161).

4. "GENIFER." Berkeley, CA: Bytel Corporation, 1986 (415-527-1157).

5. "THE KNOWLEDGEWARE CASE SOLUTION." Atlanta, GA: KnowledgeWare, 1987 (404-231-8575).

6. Brancheau, J., and J. Wetherbe. "Key Issues in Information Systems Management," *MIS Quarterly*, Volume 11, Number 1, March 1987, pp. 23–45.

7. Gane, C. *Computer-Aided Software Engineering*. New York: Rapid System Development Inc., 1988.

8. Goldstein, R. *Database: Technology and Management*. New York: John Wiley & Sons, 1985.

9. Yourdon, E. "What Ever Happened to Structured Analysis?" *Datamation*, Volume 32, Number 6, June 1986, pp. 133–138.

APPENDIX

Three Programs from the Horatio & Co. Cost Control System

Three programs from the initial prototype of the Horatio & Co. Cost Control System are contained in this appendix. They were produced by the GENIFER code generator.

Read the programs to understand their structure first. Once you have absorbed the structure of the programs, concentrate on the details of the various features. Finally, run the programs on the computer and trace the execution in the listings.

The first program is the cost control system main menu, which is labeled COST on the system structure chart, Figure 9.30. COST runs one of the programs on the first level of the system structure chart, according to the user's choice entered at the main menu screen, Figure 2.1.

All of the cost control system programs are contained on the cost control system diskette. They can be read and/or printed with any word processor capable of handling ASCII files.

```
* Menu: COST
* Generated with GENIFER
* Author: Bill Amadio with GENIFER
* Date: xx/xx/xx  Time: xx:xx
* Portions Copyright (c) 1986 Bytel Corporation. Use licensed by Bytel Corp.

set echo off
set talk off
set bell off
SET STATUS OFF
SET ESCAPE OFF
close databases
private exited
exited = .F.
do while .not. exited
  clear
  @  1, 0 say 'HORATIO & CO.'
  @  2, 0 say 'COST CONTROL SYSTEM'
  @  3, 0 say 'MAIN MENU'
  @  6,27 say '1. MAINTAIN BUDGETS'
  @  8,27 say '2. MAINTAIN EXPENSES'
  @ 10,27 say '3. INQUIRIES'
  @ 12,27 say '4. REPORTS'
  @ 14,27 say "5. ENGINEERS' ACTIVITY ANALYSIS"
  @ 16,27 say '6. PURGE JOBS'
  @ 19,27 say 'Q. QUIT'
  @ 22,27 say 'Enter choice'
  choice = ' '
  do while .not. choice $ 'Q123456'
    choice = 'Q'
    @ 22,41 get choice picture '@!'
    read
  enddo
  @ 24,00
  @ 24,00 say 'Please wait'
  do case
    case choice = 'Q'
      exited = .T.
    case choice = '1'
      if file('MB0000.prg')
        do MB0000
      else
        @ 23,00
        wait 'Menu not found...'
      endif
```

```
    case choice = '2'
      if file('ME0000.prg')
        set procedure to ME0000
        do ME0000
      else
        @ 23,00
        wait 'Procedure file not found...'
      endif
    case choice = '3'
      if file('SI0000.prg')
        do SI0000
      else
        @ 23,00
        wait 'Menu not found...'
      endif
    case choice = '4'
      if file('PR0000.prg')
        do PR0000
      else
        @ 23,00
        wait 'Menu not found...'
      endif
    case choice = '5'
      if file('EA0000.prg')
        set procedure to EA0000
        do EA0000
      else
        @ 23,00
        wait 'Procedure file not found...'
      endif
    case choice = '6'
      if file('PJ0000.prg')
       do PJ0000
      else
        @ 23,00
        wait 'Menu not found...'
      endif
  endcase
enddo (while .not. exited)
clear
SET STATUS ON
SET ESCAPE ON
SET TALK ON
SET BELL ON
return
```

The second program is the EXPENSE HISTORY REPORT BY JOB, which is labeled PR8000 on the system structure chart, Figure 9.30. The report writer specifications for the report appear in Figure 10.8. Examples of the report appear in Figures 2.10 and 3.14.

```
PROGRAM: PR8000.PRG

* Description:  EXPENSE HISTORY REPORT BY JOB
* Author:       Bill Amadio with GENIFER
* Date:         xx/xx/xx     Time: xx:xx
* Portions Copyright (c) 1986 Bytel Corporation. Use licensed by Bytel Corp.

PROCEDURE PR8000

* FILES
  database = 'EXPENSES'
  index1 = 'EXPIN3'
  secondary = 'ID'

* MEMORY VARIABLES
  page_no = 1
  line_no = 1
  record_no = 0
  top_marg = 6
  bot_marg = 6
  max_line = 66
  pg_ft_lins =  0
  outp_dev = ' '
  pr_ok = ' '
  new_sort = ' '
  files_ok = .T.
  ok = .F.
  break_on = .T.
  top_record = 0
  records = 0
  recs_on_pg = 0
  pr8000_001 = '(ACCOUNT+"-"+secondary->DESCRIPT)'
  PR8000_002 = 'CMONTH(DATE)'

* BREAK FIELDS
  job = ' '
  date = ctod('  /  /  ')

* ENVIRONMENT
  set heading off
  set scoreboard off
  set safety off
  set decimals to 10
  set margin to 0
```

```
* SCREEN
  clear
  @ 01,00 say 'PR8000.PRG'
  @ 01,16 say 'EXPENSE HISTORY REPORT BY JOB'
  @ 01,54 say 'Primary File: ' + database

* OPEN FILES
  do chk_fils
  if .not. files_ok
    return
  endif
  select 1
  use &database alias database
  select 2
  use &secondary index ref_ind alias secondary
  select database
  set relation to ACCOUNT into secondary

* INDEX
    new_sort = 'N'
  if new_sort = 'Y'
    sort_exp = 'job + date' + space(50)
    legal_sort = .F.
    do while .not. legal_sort
      @ 06,00 say 'Index expression:' get sort_exp
      read
      if type(sort_exp) = 'U' .and. '' <> trim(sort_exp)
        do disp_msg with 'Undefined sort expression'
        @ 23,00
      else
        legal_sort = .T.
        if '' <> trim(sort_exp)
          index on &sort_exp to prim_ind
        endif
      endif
    enddo
  else
    if file(index1+'.ndx')
      set index to &index1
    endif
  endif (new sort = 'Y')

* DEVICE
  do while .not. ok
    @ 23,00
    @ 23,00 say 'Printer/Screen/Quit?'
    outp_dev = ' '
```

```
   do while .not. outp_dev $ 'PSQ'
     outp_dev = 'P'
     @ 23,21 get outp_dev picture '!'
     read
   enddo
   ok = .T.
   do case
     case outp_dev = 'P'
       @ 23,24 say 'Printer ready?'
       pr_ok = ' '
       do while .not. pr_ok $ 'YN'
         pr_ok = 'Y'
         @ 23,39 get pr_ok picture '!'
         read
       enddo
       if pr_ok = 'N'
         ok = .F.
       else
         @ 23,00 say 'Printing' + space(40)
         set device to print
         set console off
       endif
     case outp_dev = 'S'
       max_line = 23
       top_marg = 0
       bot_marg = 0
       clear
     case outp_dev = 'Q'
       close databases
       return
   endcase
   enddo (while .not. ok)

* REPORT HEADER
  go top
  record_no = recno()
  line_no = top_marg
  do rept_hd
  top_record = recno()
  recs_on_pg = 0

* REPORT BODY
  do while .not. eof()
    record_no = recno()
    if job <> m->job .or. break_on
      break_on = .T.
      job = job
      condition = 'job=m->job'
```

```
        do brk1_head
        break1_rec = recno()
      endif
      if MONTH(DATE) <> MONTH(M->DATE) .or. break_on
        break_on = .T.
        date = date
        condition = 'job=m->job .and. MONTH(DATE)=MONTH(M->DATE)'
        do brk2_head
        break2_rec = recno()
      endif
      break_on = .F.
      do detail
      recs_on_pg = recs_on_pg + 1
      skip
      if job<>m->job .or. MONTH(DATE)<>MONTH(M->DATE)
        skip -1
        condition = 'job=m->job .and. MONTH(DATE)=MONTH(M->DATE)'
        do brk2_foot
        skip
      endif
      if job<>m->job
        skip -1
        condition = 'job=m->job'
        do brk1_foot
        skip
      endif
    enddo (while .not. eof())

* REPORT FOOTER
  do rept_ft
  if outp_dev = 'S'
    @ 23,79
    wait 'Press any key'
  else
    eject
    set console on
    set device to screen
  endif (outp_dev = 'S')

  close databases
return

* PROCEDURES

procedure adv_line
  line_no = line_no + 1
```

```
      if line_no > max_line - bot_marg - pg_ft_lins
         do adv_page
      endif
   return

   procedure adv_page
      page_no = page_no + 1
      if outp_dev = 'S'
         @ 23,79
         wait 'Press any key'
         clear
      else
         eject
      endif
      top_record = recno()
      recs_on_pg = 0
      line_no = top_marg
      line_no = line_no + 1
   return

   procedure brk1_head
      do adv_line
      do adv_line
      @ line_no,  0 say '** JOB'
      @ line_no,  7 say job picture '@!'
   return

   procedure brk1_foot
      do adv_line
      do adv_line
      @ line_no,  0 say '** Subtotal for job'
      go break1_rec
      sum amount while &condition to value
      go record_no
      @ line_no, 68 say value picture '9999999.99'
   return

   procedure brk2_head
      do adv_line
      do adv_line
      @ line_no,  0 say '*'
      @ line_no,  2 say &pr8000_002
   return

   procedure brk2_foot
      do adv_line
      @ line_no,  0 say '* Subtotal for month'
      go break2_rec
```

```
    sum amount while &condition to value
    go record_no
    @ line_no, 68 say value picture '9999999.99'
return

procedure chk_fils
* Return files_ok = .T. if all required files exist .F. otherwise.
* Create index file if necessary.
    close databases
    if .not. file (database + '.dbf')
       do disp_msg with 'File &database..DBF not found'
       files_ok = .F.
    endif
    if .not. file (secondary + '.dbf')
       do disp_msg with 'File &secondary..DBF not found'
       files_ok = .F.
    else
       use &secondary
       index on ACCOUNT to ref_ind
       use
    endif
return

procedure detail
* Compute and display line of information
    do adv_line
    @ line_no,  0 say date picture '@D'
    @ line_no,  9 say source picture '@!'
    @ line_no, 27 say &pr8000_001 picture '@!'
    @ line_no, 49 say descriptn picture '@!'
    @ line_no, 70 say amount picture '99999.99'
return

procedure disp_msg
parameters message
    @ 23,00
    @ 22,79
    wait message + '...'
return

procedure rept_hd
    * Print report header
    do adv_line
    @ line_no,  0 say 'DATE:'
    @ line_no,  6 say date()
    @ line_no, 69 say 'PAGE:'
    @ line_no, 75 say page_no picture '999'
```

```
   do adv_line
   @ line_no,  0 say 'TIME:'
   @ line_no,  8 say substr(time(),1,5)
   do adv_line
   @ line_no, 33 say 'HORATIO & CO.'
   do adv_line
   @ line_no, 30 say 'COST CONTROL SYSTEM'
   do adv_line
   @ line_no, 25 say 'EXPENSE HISTORY REPORT BY JOB'
   do adv_line
   do adv_line
   do adv_line
   @ line_no,  0 say 'DATE       SOURCE              ACCOUNT'
   @ line_no, 49 say 'DESCRIPTION            AMOUNT'
   do adv_line
return

procedure rept_ft
   * Print report footer
   do adv_line
   do adv_line
   do adv_line
   @ line_no,  0 say '*** Grand total'
   sum amount to value
   IF RECORD_NO<=RECOUNT()
     go record_no
   ENDIF
   @ line_no, 68 say value picture '9999999.99'
return

* EOF PR8000.PRG
```

The final program is the maintenance program for the file BUDGET, which is labeled MB2000 on the system structure chart, Figure 9.30. The data dictionary specifications for BUDGET appear in Figure 10.3. The program is long due to the comprehensive list of options that the GENIFER maintenance programs present. See Figure 3.4.

```
* PROGRAM: MB2000.PRG

* Description:   MAINTAIN MONTHLY BUDGET FIGURE
* Author:        Bill Amadio with GENIFER
* Date:          xx/xx/xx    Time: xx:xx
* Portions Copyright (c) 1986 Bytel Corporation. Use licensed by Bytel Corp.

* MAINTENANCE OPTIONS:
   * Duplic. keys: no
   * Modify key:    yes
   * Auto add:      yes
   * Partial key:   yes
   * Comments:      yes
   * Help screens: yes

PROCEDURE MB2000

* FILES
   * MONTHLY BUDGET FIGURES
   database1 = 'BUDGET'
   index_fil1 = 'BUDIN1'
   ref_fil1 = 'ID'
   ref_idx1 = 'IDIN1'

* VARIABLES
   * Key in database BUDGET
     account = ''
     job = ''
     me_date = ctod('  /  /  ')
   * Fields in database BUDGET
     bud_mnth = 0

   * Flags
     abort_rec = .F.
     dupl_rec = .F.
     empty = .F.
     files_ok = .T.
     del_rec = .F.
     valid_rec = .T.
     show_all = .T.
     filter_on1 = .F.
     set_filter = .F.
```

```
  * Others
    key1 = 'm->account + m->job + str(year(m->me_date),4) + str(month
(m->me_date),2) + str(day(m->me_date),2)'
    index1 = 'account + job + str(year(me_date),4) + str(month(me_date),2)
+ str(day(me_date),2)'
    null_key1 = '            0 0 0'
comp_key1 = 'trim(m->account + m->job + str(year(m->me_date),4) + str(month
(m->me_date),2) + str(day(m->me_date),2))'

    list1 = 'account,job,me_date,bud_mnth'
    pak1 = .F.
    choice = ''
    option = ''
    filt_str = ''
    filter1 = ''
    scr_num = 1
    num = '1'
    max_screen = 1
    record_no = 0


* ENVIRONMENT
  set exact off
  set heading off
  set bell off
  set deleted on

* SCREEN
  do disp_scr

* OPEN FILES
  do chk_fils
  if .not. files_ok
    return
  endif
  index_str = index_fil1
  select 1
  use &database1 index &index_str
  select 2
  use ID index &ref_idx1 alias ref_fil1
  select budget
  SET RELATION TO ACCOUNT INTO REF_FIL1
  record_no = recno()
  * database is empty if eof() = .T.
  empty = eof()
  if empty
    do disp_msg with 'Database empty'
  endif
```

```
* PROCESSING LOOP
  do while scr_num >= 1
    record_no = recno()
    * preserve option if in add mode
    if empty .or. (scr_num = max_screen .and. option = 'A')
      option = 'A'
    else
      do load_var
      do disp_rec
      option = 'M'
    endif
    do get_optn with 'Ret/Beg/End/Next/Prev/Skip/Modify/Add/Copy/' + ;
                     'Del/List/Filt/Tally/Help/Quit','RBENPSMACDLFTHQ',option

    do case
    * Add
      case option = 'A'
        do add
    * Beginning
      case option = 'B'
        go top
    * Copy
      case option = 'C'
        do add
    * Delete
      case option = 'D'
        do delete
    * End
      case option = 'E'
        go bottom
    * Filter
      case option = 'F'
        do filter
    * Help
      case option = 'H'
        do help
        show_all = .T.
        do disp_scr
    * List
      case option = 'L'
        do list with list&num
        show_all = .T.
        do disp_scr
        go record_no
    * Modify
      case option = 'M'
        do modify
```

```
    * Next
      case option = 'N'
        skip
        if eof()
          go bottom
          do disp_msg with 'Last record'
        endif
    * Previous
      case option = 'P'
        skip -1
        if bof()
          go top
          do disp_msg with 'First record'
        endif
    * Retrieve
      case option = 'R'
        do retrieve
    * Skip
      case option = 'S'
        do skip
    * Tally
      case option = 'T'
        do tally
      endcase

  * Quit
    if option = 'Q'
      scr_num = scr_num - 1
      num = str(scr_num,1)
      if scr_num <> 0
        select &num
        empty = .F.
        show_all = .T.
        do disp_scr
      endif
    endif
  enddo (while scr_num >= 1)

  if del_rec
    do pack_all
  endif
  close databases
return
```

```
* PROCEDURES (listed alphabetically)

procedure add
* Add a new record (if option = 'A') or copy the current
* record (if option = 'C')
  if option = 'A'
    do init_key
    do init_fld
    do clr_cal
  endif
  * The following "get record" loop can be exited either when a valid
  * record is entered (valid_rec = .T.), or when entry is aborted by
  * (1) blank key, (2) duplicate key (if not allowed), or (3) abort
  * request from the validation procedure (if called).
  abort_rec = .F.
  valid_rec = .F.
  do while .not. (valid_rec .or. abort_rec)
    do get_key
    do get_flds
    read
    comp_key = comp_key&num
    if null_key&num = &comp_key
      * blank key
      abort_rec = .T.
    else
      do chk_dupl
      if dupl_rec
        abort_rec = .T.
      else
*         do disp_cal
        do val_rec
      endif (dupl_rec)
    endif (null_key = comp_key)
  enddo (while .not. (valid_rec .or. abort_rec))
  if valid_rec
    do save_rec with .T.
  else
    * break out of "Add" if record invalid
    if empty
      option = 'Q'
    else
      option = 'M'
      if .not. dupl_rec
        go record_no
      endif
    endif (empty)
  endif (valid_rec)
return
```

```
procedure chk_dupl
* Set dupl_rec to .F. if key is a duplicate, or to .T. otherwise
  key = key&num
  if .not. empty
    seek &key
  endif
  if option = 'M'
    dupl_rec = record_no <> recno() .and. .not. eof()
  else
    dupl_rec = .not. eof()
  endif
  if dupl_rec
    do disp_msg with chr(7) + 'Duplicate key not allowed'
  else
    if .not. empty
      go record_no
    endif
  endif dupl_rec
return

procedure chk_fils
* Set files_ok to .F. and display a message if a file is missing;
* create an index file if one does not exist.
  close databases
  file_num = 1
  do while file_num <= max_screen .and. files_ok
    seq = str(file_num,1)
    if .not. file (database&seq + '.dbf')
      do disp_msg with 'File ' + database&seq + ' not found'
      files_ok = .F.
    endif
    if files_ok .and. .not. file (index_fil&seq + '.ndx')
      @ 23,00
      @ 23,00 say 'Indexing'
      db = database&seq
      ind = index_fil&seq
      use &db
      ind_key = index&seq
      index on &ind_key to &ind
      use
    endif
  file_num = file_num + 1
  enddo (while file_num <= max_screen .and. files_ok)
  if .not. file ('ID' + '.dbf')
    do disp_msg with 'File ID not found'
    files_ok = .F.
```

```
      else
        if .not. file ('&ref_idx1' + '.ndx')
          @ 23,00
          @ 23,00 say 'Indexing'
          use ID
          index on ACCOUNT to &ref_idx1
        endif
      endif
      @ 23,00
      @ 24,00
return

procedure clr_cal
* Clear computed screen areas
   @  8,45 say space (30)
return

procedure clr_flds
* Clear screen field areas
   @ 19, 8 say space (8)
return

procedure delete
* Delete current record upon user approval
   choice = 'N'
   do get_optn with 'Delete (Y/N)','YN',choice
   if choice = 'Y'
     delete
     * reposition to the next record
     skip
     * if last record deleted, go to beginning of database
     if eof()
       go top
     endif
     if eof() .and. filter_on&num
       * if no records left in filter, remove the filter
       filt_str = filter&num
       set filter to &filt_str
       filter_on&num = .F.
       @ 0,0 say space(9)
       go top
     endif (eof() .and. filter_on&num)
     if eof()
       * quit if last record deleted and database becomes empty
       option = 'Q'
     endif
```

```
      del_rec = .T.
      pak&num = .T.
   endif (choice = 'Y')
return

procedure disp_cal
* Evaluate and display computed fields
   mb2000_001 = REF_FIL1->DESCRIPT
   @  8,45 say mb2000_001 picture '@!'
return

procedure disp_msg
parameters message
* Display "message" + '...' at line 23; wait for entry of any key
   @ 23,00
   @ 22,79
   wait message + '...'
return

procedure disp_rec
* Load fields from database and display record
   do get_key
   do get_flds
   clear gets
   do disp_cal
   show_all = .F.
return

procedure disp_scr
* Display stationary part of screen
   clear
   @  1, 0 say 'HORATIO & CO.'
   @  2, 0 say 'COST CONTROL SYSTEM'
   @  3, 0 say 'MAINTAIN MONTHLY BUDGET FIGURES'
   @  7, 0 say 'Account                           `Accou'
   @  7,40 say 'nt'
   @  8, 0 say 'Number                             Name'
   @ 11, 0 say 'Job'
   @ 14, 0 say 'Month'
   @ 15, 0 say 'Ending Date'
   @ 18, 0 say 'Budget'
   @ 19, 0 say 'Amount'
   if filter_on&num
      @ 0,0 say 'FILTER ON'
   else
      @ 0,0 say space(9)
   endif
return
```

```
procedure filter
* Set filter on database
  choice = 'Y'
  do get_optn with 'Set Filter (Yes/No/Cancel)','YNC',choice
  if choice = 'Y'
    set_filter = .T.
    do init_key
    do init_fld
    do clr_cal
    do get_key
    do get_flds
    read
    filt_str = ''
    if '' <> trim(m->account)
      filt_str = filt_str + 'account = trim("&account") .and.'
    endif
    if '' <> trim(m->job)
      filt_str = filt_str + 'job = trim("&job") .and.'
    endif
    if .not. (m->me_date <> m->me_date .or. .not. (m->me_date = m->me_date))
      me_date = dtoc(m->me_date)
      filt_str = filt_str + 'me_date = ctod("&me_date") .and.'
    endif
    if m->bud_mnth <> 0
      bud_mnth = str(m->bud_mnth,8,2)
      filt_str = filt_str + 'bud_mnth = &bud_mnth .and.'
    endif
    if '' = trim(filt_str)
      filt_str = filter&num
      set filter to &filt_str
      filter_on&num = .F.
    else
      if '' <> trim(filter&num)
        filt_str = filter&num + '.and.' + filt_str
      endif
      filt_str = substr(filt_str,1,len(filt_str)-6)
      set filter to &filt_str
      go top
      filter_on&num = .not. eof()
      if .not. filter_on&num
        do disp_msg with 'No records match the filter'
        filt_str = filter&num
        set filter to &filt_str
        go record_no
      endif
    endif ('' = trim(filt_str))
  endif (choice = 'Y')
```

```
  if choice = 'C'
    filt_str = filter&num
    set filter to &filt_str
    filter_on&num = .F.
  endif (choice = 'C')
  if filter_on&num
    @ 0,0 say 'FILTER ON'
  else
    @ 0,0 say space(9)
  endif
  set_filter = .F.
return

procedure get_flds
* Get field variables
  @ 19, 8 get m->bud_mnth picture '99999.99'
return

procedure get_key
* Get key variables
  @  8, 9 get m->account picture '@9'
  @ 11, 5 get m->job picture '@!'
  @ 15,13 get m->me_date picture '@D'
return

procedure get_optn
parameters message, choices, choice
* Display the string "message" on line 23; get a character
* (defaulted to "choice"), validate it against "choices" and
* return in "choice" (must be a memory variable, not literal).
* If choice = 'A' display choice without accepting it.
  @ 23,00 say message + '? '
  @ 23,len(message)+3
  char = ' '
  do while .not. char $ choices
    char = choice
    @ 23,len(message) + 2 get char picture '!'
    if choice <> 'A'
      read
    else
      clear gets
    endif
  enddo (while .not. char $ choices)
  choice = char
return
```

```
procedure help
* Display help screens
  clear
  @ 03,00
  text
                  Option    Database maintenance action

                  _____    _____
                  Add       add a record to the database
                  Beg       go to beginning of database
                  Copy      duplicate current record
                  Del       delete current record
                  End       go to end of database
                  Filt      set filter on database
                  List      display records on screen
                  Modify    edit current record
                  Next      go to next record
                  Prev      go to previous record
                  Quit      terminate current activity
                  Ret       retrieve a record by key
                  Skip      move up or down by a specified number of records
                  Tally     count records
  endtext
  do disp_msg with 'More'
  clear
  @ 05,00
  text
                  White key    Screen editing action

                  _____    _____
                  -->          character right
                  <--          character left
                  up arrow     previous field
                  down arrow   next field
                  PgDn, PgUp   accept screen
                  End          next word/field
                  Home         previous word/field
                  Del          delete character
                  Ins          insert on/off toggle
  endtext
  do disp_msg with 'OK'
  clear
return

procedure init_fld
* Clear field variables
  bud_mnth = 0.00
return
```

```
procedure init_key
* Clear key variables
  account = space(4)
  job = space(4)
  me_date = ctod('  /  /  ')
return

procedure list
parameters list_items
* List records beginning from current record
  do while .T.
    clear
    list off next 20 &list_items
    if eof()
      do disp_msg with 'OK'
      exit
    endif
    choice = 'Y'
    do get_optn with 'More','YN',choice
    if choice = 'N'
      exit
    endif
  enddo (while .T.)
return

procedure load_var
* Copy fields from database record to memory variables
  account = account
  job = job
  me_date = me_date
  bud_mnth = bud_mnth
return

procedure modify
* Modify current record
* The following "get record" loop is exited when either (1) a valid record
* is entered (valid_rec = .T.), or (2) entry is aborted by a duplicate
* key (when not allowed), or a request from the validation procedure.
  valid_rec = .F.
  abort_rec = .F.
  do while .not. (valid_rec .or. abort_rec)
    do get_key
    do get_flds
    read
    comp_key = comp_key&num
    if null_key&num = &comp_key
      abort_rec = .T.
```

```
      else
        do chk_dupl
        if dupl_rec
          abort_rec = .T.
        else
*            do disp_cal
          do val_rec
        endif (dupl_rec)
      endif (null_key = comp_key)
    enddo (while .not. (valid_rec .or. abort_rec))
    if valid_rec
      do save_rec with .F.
    else
      if .not. dupl_rec
        go record_no
      endif
      go record_no
    endif (valid_rec)
return

procedure pack_all
* Pack deleted records with user confirmation
    choice = 'N'
    do get_optn with 'Reorganize database? (Y/N)','YN',choice
    if choice = 'Y'
      @ 23,00
      @ 23,00 say 'Packing, please wait'
      do while scr_num < max_screen
        scr_num = scr_num + 1
        num = str(scr_num,1)
        if pak&num
          select &num
          pack
        endif
      enddo (while scr_num < max_screen)
    endif
return

procedure repl_rec
* Replace database fields with memory variables
    replace account with m->account
    replace job with m->job
    replace me_date with m->me_date
    replace bud_mnth with m->bud_mnth
return
```

```
procedure retrieve
* Accept key and seek record; if not found reposition to record_no
  do clr_flds
  do clr_cal
  do init_key
  do get_key
  read
  comp_key = comp_key&num
  if null_key&num = &comp_key
    * blank key
    return
  endif
  key = key&num
  seek trim(&key)
  if eof()
    do disp_msg with 'Not found'
    go record_no
  endif
return

procedure save_rec
parameters new_rec
* If new_rec: append record currently in memory to database;
* if .not. new_rec: replace database record record_no with memory fields.
  choice = 'Y'
  do get_optn with 'Save (Y/N)','YN',choice
  if choice = 'Y'
    if new_rec
      append blank
    endif
    do repl_rec
    empty = .F.
    if filter_on&num
      * Check to see if record matches filter
      record_no = recno()
      skip
      skip -1
      if recno() <> record_no
        go top
        if eof()
          * Remove filter
          filt_str = filter&num
          set filter to &filt_str
          filter_on&num = .F.
          @ 0,10 say space(9)
          go record_no
        endif
      endif
    endif (filter_on&num)
```

```
    else
      if .not. empty
        go record_no
      endif
    endif (choice = 'Y')
return

procedure skip
* Move forward/backward several records
  skip_no = 0
  @ 23,70 say '  Recs' get skip_no picture '@Z 999'
  read
  skip skip_no
  if eof()
    go bottom
    do load_var
    do disp_rec
    do disp_msg with 'Last Record'
  endif
  if bof()
    go top
    do load_var
    do disp_rec
    do disp_msg with 'First record'
  endif
return

procedure tally
* Count and display number of records in database
  @ 23,00
  @ 23,00 say 'Counting, please wait'
  count to number_
  do disp_msg with 'Count: ' + str (m->number_,6) + ' records'
  go record_no
return

procedure val_rec
* Set valid_rec to .F. if a field doesn't pass validation test;
* set abort_rec to .T. or .F., depending on user's response
  valid_rec = .T.
  select ref_fil1
  seek m->account
  if eof()
    valid_rec = .F.
    ?? chr(7)
    choice = 'Y'
    do get_optn with "ACCOUNT NUMBER NOT FOUND IN ID; display (Y/N)",'YN',choice
```

```
   if choice = 'Y'
     go top
     if type ('descript') <> 'U'
       do list with 'ACCOUNT,DESCRIPT'
     else
       do list with 'ACCOUNT'
     endif
     select budget
     show_all = .T.
     do disp_scr
     do disp_rec
   endif
 ELSE
   DO DISP_CAL
 endif
 select budget
 abort_rec = .F.
 if .not. valid_rec
   choice = 'Y'
   do get_optn with 'Invalid entries.  Correct (Y/N)', 'YN', choice
   if choice = 'N'
     abort_rec = .T.
   endif
 endif (.not. valid_rec)
return

* EOF MB2000.PRG
```

GLOSSARY

access paths   The ways in which a system's programs access the system's data.

artificial intelligence (AI)   Efforts to expand computer-based technology into more sophisticated mental processes such as problem solving, pattern recognition, and natural language processing.

asynchronous communication   A development brought about by technologies such as **computer conferencing**. All participants in a conference do not have to be involved at the same time.

audioconferencing   Technology that provides sound communication among geographically dispersed participants.

audit trail   Documents that provide the ability to determine the activity represented by the values in the master data store status fields of a transaction processing system.

automatic code generation   A capability of many CASE systems. Code generator programs write application programs from user specifications.

background information   Identification information about system entities stored in the master data store of a transaction processing and/or reporting system. For example, a customer's name and address in an accounts receivable system. See also **status information**.

backup   The process of making safekeeping copies of a system's data and programs. Backup copies are used to restore the system in the event of a disaster such as a fire.

batch control   A mechanism to control the accuracy of transaction maintenance data entry. Computer-generated sums and counts are compared to manually generated figures to identify the presence of errors.

boundary of the new system   The processes of the Analysis DFD included in the Requirements Model and the new processes specified in the Requirements Model.

break footer   The portion of a control-break report that is printed at the end of a group. Group subtotals are printed in the break footer section.

break header   The portion of a control-break report that is printed at the start of a new group. Group identification is usually printed in the break header section.

browsing mechanism A system feature that allows the user to move through a database in search of a specific record by displaying groups of database records on the screen.

business objectives Broadly stated, measurable outcome used to guide the activities and decisions of an enterprise.

business tactics Specific actions performed to achieve business objectives.

buy/build software phase The phase of the systems development life cycle concerned with the programming or acquisition of the software component.

centralized processing environment An environment in which all users access the same computer and database.

command level processing A way of using an interactive database management system. Commands entered from the keyboard are executed immediately, and the results are displayed on the screen or printer.

components Individual data items in a contents form.

computer-assisted design (CAD) Graphics and text-oriented computer and communication systems used to improve the efficiency and effectiveness of design engineers.

computer-assisted manufacturing (CAM) Automated systems that support the manufacturing function. Applications include the storage and retrieval of inventory, the movement of materials within the plant, automatic assembly, and test equipment.

computer conferencing Technology that enables geographically dispersed participants to interact using computer terminals.

computer-integrated manufacturing (CIM) A term applied to projects aimed at integrating the business, engineering, and manufacturing systems of a firm.

computer programmers Professionals who write the instructions that direct the computer to perform its tasks.

computer-aided software engineering (CASE) Tools to automate systems development tasks such as program coding, documentation management, problem analysis, and systems design.

computer-based knowledge The storage of data along with the rules to use the data for problem solving. See also **inference engine**.

conservation of data flows When expanding a data flow diagram process to a new level, one should not add or delete data flows that cross the boundary of the process in question.

contents form A device used in database design to record the contents of new system data flows and data stores.

context data flow diagram A high-level data flow diagram showing data flows that cross the boundary of the system being modeled.

context-sensitive help Computer-generated operator instructions/support callable by the user at the point of need.

control fields The field(s) of the input file record used to control grouping in a control-break report.

control information System-wide information used by several modules of a system. For example, the current date or the name of the company.

control mechanism A device used to detect substandard performance.

control-break report A columnar report that usually prints one line for every input record. The distinguishing characteristic is the grouping of records according to the values of one of the fields of the record. Summary information such as subtotals is often printed whenever the value of the control field changes or breaks.

cycle of transaction maintenance-update-reports The operating cycle of a transaction processing system. Transactions are captured, their effect is recorded in the master, and information is returned to the user.

data capture The transaction processing function that identifies and records day-to-day transaction activity.

data dictionary A collection of information about the data items maintained in files and databases. In modern environments, data dictionaries extend beyond data items to reports, screens, and system entities.

data flow diagram (DFD) A systems analysis model that represents the processing activities of a system.

data flows The paths upon which information and/or objects move in a data flow diagram.

data store Generally, a place where data is stored. In computer-based systems a data store is usually a file or a group of related files.

data structure diagram A graphical model used to express relationships among data stores and files.

data validation The process of checking keyboard entries against a file of valid items for a match.

data verification The process of displaying information to support accurate data entry. For example, displaying the name of an inventory part when the user enters a part number.

data-oriented modeling Systems development models that focus on the data requirements of a system. For example, entity-relationship models.

day-to-day activity Activity represented by the transactions of a transaction processing system.

decision support systems (DSS) Flexible systems that support less-structured and less-repetitive management activities.

design database Typically holds information about the data to be maintained by a system, the business logic of the processes to be implemented, the physical layouts of the screens and reports, and other design information.

design DFD A logical model of the automated portion of the new system. The Design DFD shows the central data stores of the new system, the relationships among the data stores, and the potential sources of the data stores' contents.

design new system phase Design activities of the systems development life cycle. Components include hardware, software, data, procedures, and personnel.

designer's tradeoff chart A decision aid used to determine a suitable mix of system features, responsibilities, and costs.

detail A term commonly applied to the "many" file of a one-to-many relationship.

detail lines Lines comprising the body of a report.

discovery prototyping Prototyping used during the early stages of the life cycle to help with the initial determination of information requirements. See also **refinement prototyping.**

distributed processing environment An extended environment consisting of multiple computers and databases. See also **centralized processing environment**.

documentation A set of materials consisting of system operating instructions and reference documents regarding systems analysis, design, and programming.

economic feasibility The match between system costs and system benefits.

edit list In transaction maintenance processes, a list of recently entered transactions that are compared to the corresponding source documents for accuracy. See also **proof list**.

electronic data processing (EDP) An early term commonly applied to computer-based transaction processing and accounting systems.

electronic mail Systems that provide for the electronic storage and transmission of messages and documents.

end-user computing (EUC) Computer-based applications developed by the members of the user/management group with assistance from information systems professionals.

entity types The "things" of interest to a business and/or a system that need to have data stored about them. For example, customers, inventory items, etc.

entity-relationship (E-R) model A systems analysis model that represents the data requirements of a system.

evaluation of solution alternatives Study of ways to implement new system requirements.

exception report A report that scans a large database to identify extraordinary items. For example, inventory items below reorder point.

expert systems Technology that stores and uses knowledge in the form of data and the rules to use the data to solve problems.

factory automation Computer-based technology applied to the factory. Applications include robotics, computer-aided design, and production scheduling and control.

file index A separate file connected to a data file. When under the control of an index, data file records appear to be in the sequence defined by the index rather than in the actual physical sequence.

file relation The combination of two or more files in a relational database management system. The combination appears as a single entity to the user or programmer.

file superstructures In a data structure diagram, the means of expressing a data store as a collection of files.

file transfer utilities Software packages that allow the electronic transfer of data between computer systems.

final analysis DFD A data flow diagram drawn by combining the data flows of the Context DFD, the processes of the Level 0 Analysis DFD, and the data flows and the data stores of the Expansion DFDs into one diagram.

fourth-generation environment A software environment providing a relational database management system, nonprocedural programming, and other application generation tools.

function-oriented modeling Systems development models that focus on the processing requirements of a system. For example, data flow diagrams.

gantt chart A project scheduling device. Activities are assigned to individuals or groups and placed along a time line according to the required completion dates.

header A term commonly applied to the "one" file of a one-to-many relationship.

implement and evaluate new system phase The phase of the systems development life cycle during which the system is put into production and evaluated against the original requirements.

inference engine The software that allows an expert system to make conclusions based upon the knowledge stored in the knowledge base.

information architecture A high-level model of the information requirements of an organization. The model is independent of personnel, organizational structure, and technology.

information centers Centers staffed by systems specialists who provide support for the development of end-user applications.

information system (IS) A term used to represent the expansion of computer-based technology beyond transaction processing to fixed-format inquiries and reports. See also **management information system**.

initial data conversion The process of converting existing data to the format required by a new system. For example, manual records to computer files.

inquiry Similar to a report, but the data items usually concern a specific entity.

intangible benefits Benefits that are difficult to quantify without experience.

islands of automation A term used to describe applications developed in isolation without regard to interconnectivity and support for other areas of the business.

key The component or components that uniquely identify a given record in a data store.

knowledge engineer A professional involved in the analysis, design, implementation, and maintenance of expert systems.

level 0 analysis DFD A logical model of the current system drawn by separating the data flows of the Context Data Flow Diagram into logical groups, each group representing a system process.

level 1 data flow diagram A detailed expansion of one of the system processes shown in a level 0 data flow diagram.

level 2, level 3 data flow diagrams The continuing expansion of system processes shown on higher-level data flow diagrams.

leveling The process of expanding a data flow diagram into increasing levels of detail.

logical model A representation of WHAT a system does. Details of HOW the work is done are not shown. See also **physical model**.

maintain and enhance system phase The phase of the systems development life cycle representing the useful life of a system. Changes are incorporated as errors are discovered or new requirements evolve.

management information system (MIS) A term used to represent the expansion of computer-based technology beyond transaction processing to fixed-format inquiries and reports. See also **information system**.

master The file or database that contains the background and status information about transaction processing system entities.

master maintenance The process of maintaining master data store records in a transaction processing and/or reporting system. Typical functions include add, change, delete, inquire, and list.

master/transaction structure   The file structure of a transaction processing and/or reporting system. Background and status information about system entities is stored in the master; day-to-day activity is captured in the transactions.

menu   A program that presents a list of system function options to the user, accepts the user's choice, and executes the appropriate programs.

mouse   A cursor control device used for indicating menu choices and free-style drawing.

nonprocedural language   A programming language that does not require the specification of processing steps. The processing structure is assumed, and the programmer specifies his or her options for various features.

normalizing the database   The process of transforming preliminary contents forms into an effective database design by removing anomalies such as repeating groups and partial and transitive dependencies.

office automation (OA)   Office productivity and communication systems such as word processing, electronic mail, electronic conferencing, and image storage.

one-to-many relationship   A relation between files. For each value of the linking field, one file contains one matching record, and the other contains many matching records.

operational feasibility   The match between the features and responsibilities of a system and the people in the user/management group.

operational prototype   The prototype version that the user/designer feels meets his or her requirements.

page footer   The portion of a report that is printed at the bottom of each page.

page header   The portion of a report that is printed at the top of each page.

partial dependency   Data store components that do not depend upon the entire key.

performance efficiency   A measure of the output produced from a given set of inputs.

performance quality   A measure of how well a job is done.

physical model   A representation of HOW the work of a system is carried out. See also **logical model**.

problem and/or opportunity analysis phase   First phase of the systems development life cycle. Problems and/or opportunities are analyzed in regard to the usefulness of an information systems solution.

procedural language   A programming language in which processing logic is specified in a series of steps. For example, COBOL.

processes   Units of activity in a data flow diagram.

processing control   The ability to detect and prevent system processing irregularities such as keyboard errors, incorrect use of system functions, and deliberate abuse.

proof list   In transaction maintenance processes, a list of recently entered transactions that are compared to the corresponding source documents for accuracy. See also **edit list**.

prototyping   A systems development methodology based upon the quick development of a working model to stimulate user feedback.

pull-down menus   A feature of modern graphics-oriented user interfaces. A list of choices appears when needed and disappears after a choice has been made.

purging The process of removing obsolete records from a database.

record retrieval by key value The process of accessing a record in a database through the specification of its unique identification key value. For example, checking a credit card balance through the account number.

redundant nonkey components Components that appear in more than one data store and that are not part of the key of any data store.

reference data store The data store that holds the information used to check keyboard entries during data validation processes.

refinement prototyping Basic information requirements are established before the prototyping process is applied to the development of the software and data components of the system. See also **discovery prototyping.**

relationship types These exist between entity types. They are described by the role each entity type plays in the relationship and the number of entity types that can participate in the relationship. For example, one representative services many customers.

reorganization A rebuilding of a database to improve processing efficiency. Activities include removing data records marked for deletion and reconstructing file indexes.

repetition A component in a contents form that takes on a series of values. For example, monthly budgets.

report A listing of data items organized into a predetermined format.

report footer The portion of a report that is printed once at the end of the report.

report header The portion of a report that is printed once at the beginning of the report.

report writer A software tool that automatically generates reports or report programs from user specifications.

requirements model A statement of new system functional requirements based upon an analysis of business tactics and system objectives, along with data, access, and control issues.

restore Using backups to set a system in operation after a disaster such as a fire.

reverse engineering The process of analyzing the programs of an existing system to generate the system design database automatically.

screen painters Software tools that generate menu and file maintenance programs from user specifications.

semantics The meaning behind descriptive terms used in systems modeling and design. For example, "special orders."

sequence The order in which records are stored or displayed.

source documents Manual documents that serve as the source of computer system data entry.

status information Information, stored in the master data store of a transaction processing system, that is updated by transactions. For example, a customer's outstanding balance in an accounts receivable system.

steering committee A representative group of organization members charged with the responsibility of planning, organizing, and controlling information system resources.

structured English A pseudocode-like device used to record system processing specifications.

summarization   A common way to transform vast amounts of data into useful information. For example, the cost control system's Job Cost Reports.

system   An entity that transforms inputs into outputs.

system builder   One-half of a typical prototyping project team. The system builder assumes the more technical role of constructing a working system that satisfies the user/designer's requirements. See also **user/designer**.

system objectives   Broadly stated system actions developed to support a set of business tactics.

system structure chart   A graphical model used to illustrate the contents of system processes and the hierarchical relationships among the processes.

systems analysis and design   The activities undertaken by members of a business organization in the development and maintenance of computer-based information systems.

systems analysis model building phase   Study of the current means of addressing problem and/or opportunity and the determination of new system requirements.

systems analysts   Professionals who assist in the identification of business problems and opportunities and in the specification of information system solutions.

systems designers   Professionals who transform information system requirements into technical specifications for programming.

systems development life cycle   Systems development and maintenance activities organized into a set of phases and used as a guide for systems development projects.

tangible benefits   Benefits that are easy to quantify without experience.

team review   In a systems development project, a meeting between users and builders to review prior decisions and activities and to plan for the future.

technical feasibility   The ability of the organization's technical resources to support a given system.

telecommunications   A collection of technologies involving voice, data, and video communications.

teleconferencing   A general term applied to technology that enables people to communicate more effectively. See also **videoconferencing**, **audioconferencing**, **computer conferencing**.

terminal emulation   Software/hardware combinations that allow microcomputers to serve as minicomputer and mainframe terminals.

terminators   The entities outside the system being modeled in a data flow diagram.

third-generation language   A procedural programming language such as COBOL, PL/1, FORTRAN, and BASIC.

tradeoff   The most important word in the systems analyst's vocabulary. Finding a suitable mix of system features, responsibilities, and cost.

transaction   A unit of activity in a transaction processing system. For example, a sale, an inventory receipt, a general ledger journal entry.

transaction maintenance   The process of recording transactions in a transaction processing system. Typical functions include Add, Change, Delete, Inquire, and List.

transaction processing system   A system that maintains background and status information about entities and that records the day-to-day activity of these entities.

transitive dependency Data store components that depend upon nonkey components.

updating In a transaction processing system, the process of recording the effect of transactions on the background and status information of the master.

user interface The combination of the functions provided by a system and the format in which they are presented to the user.

user-friendly software Software tools that are accessible to users other than information systems professionals.

user/designer One-half of a typical prototyping project team. The user/designer decides upon front-end system requirements and controls the design process. See also **system builder**.

user/management group Group composed of business organization members who use a particular computer-based information system and their managers.

videoconferencing Technology that utilizes full-motion or still pictures to provide sight and sound communication among geographically dispersed participants.

voice mail Systems that provide for the electronic storage and transmission of spoken messages.

windows A feature of modern graphics-oriented user interfaces. Multiple displays appear in various places on a single screen.

INDEX

Note: Page numbers in boldface denote glossary terms.